THE EFFECTS OF SOUND ON PEOPLE

Wiley Series in Acoustics, Noise and Vibration Series list

THE EFFECTS OF SOUND ON PEOPLE

James P. Cowan

This edition first published 2016
© 2016 John Wiley & Sons, Ltd

Registered Office
John Wiley & Sons, Ltd, The Atrium, Southern Gate, Chichester, West Sussex, PO19 8SQ, United Kingdom

For details of our global editorial offices, for customer services and for information about how to apply for permission to reuse the copyright material in this book please see our website at www.wiley.com.

Library of Congress Cataloging-in-Publication Data

Names: Cowan, James P., author.
Title: The effects of sound on people / James P. Cowan.
Description: Chichester, West Sussex, United Kingdom : John Wiley & Sons, Inc., [2016] |
   Includes bibliographical references and index.
Identifiers: LCCN 2016003038| ISBN 9781118895702 (cloth) | ISBN 9781118895689 (epub) |
   ISBN 9781118895672 (Adobe PDF)
Subjects: LCSH: Hearing. | Sound–Physiological effect.
Classification: LCC QP461 .C69 2016 | DDC 612.8/5–dc23
LC record available at http://lccn.loc.gov/2016003038

A catalogue record for this book is available from the British Library.

Set in 10/12.5pt Times by SPi Global, Pondicherry, India
Printed and bound in Singapore by Markono Print Media Pte Ltd

1   2016

To Al, Lorraine, Lynn, Josh, Beth, and Noah.

Contents

List of Figures

List of Tables

About the Author

James P. Cowan is a board-certified noise control engineer with more than 30 years' experience in noise control, architectural acoustics, and environmental noise issues. He has acted as a consultant to public agencies, architects, engineers, industrial personnel, and attorneys in all areas of noise control; hearing damage and protection criteria; and acoustic design of all types of spaces. Mr. Cowan has lectured on acoustical topics to thousands of professionals, delivering live seminars and webinars, and teaching courses for universities, professional societies, and private organizations across the US for more than 25 years. In addition to several book chapters and many published articles, Mr. Cowan is the author of *Architectural Acoustics Design Guide*, published by McGraw-Hill in 2000; *Architectural Acoustics*, an interactive educational CD set published by McGraw-Hill in 1999; and *Handbook of Environmental Acoustics*, a reference book in community noise issues published by Van Nostrand Reinhold (subsequently Wiley) in 1994. He is currently Principal Acoustical Engineer at AECOM in Manchester, NH, USA and Instructor in Acoustics at the Boston Architectural College in Boston, MA, USA.

Series Preface

This book series will embrace a wide spectrum of acoustics, noise and vibration topics from theoretical foundations to real world applications. Individual volumes included will range from specialist works of science to advanced undergraduate and graduate student texts. Books in the series will review the scientific principles of acoustics, describe special research studies and discuss solutions for noise and vibration problems in communities, industry and transportation.

The first books in the series include those on biomedical ultrasound, effects of sound on people, engineering acoustics, noise and vibration control, environmental noise management, sound intensity and wind farm noise – books on a wide variety of related topics.

The books I have edited for Wiley, *Encyclopedia of Acoustics* (1997), *Handbook of Acoustics* (1998) and *Handbook of Noise and Vibration Control* (2007), included over 400 chapters written by different authors. Each author had to restrict the chapter length on their special topics to no more than about 10 pages. The books in the current series will allow authors to provide much more in-depth coverage of their topic.

The series will be of interest to senior undergraduate and graduate students, consultants, and researchers in acoustics, noise and vibration and, in particular, those involved in engineering and scientific fields, including aerospace, automotive, biomedical, civil/structural, electrical, environmental, industrial, materials, naval architecture, and mechanical systems. In addition, the books will be of interest to practitioners and researchers in fields such as audiology, architecture, the environment, physics, signal processing, and speech.

Malcolm J. Crocker
Series editor

Preface

Sound is and always has been a source of pleasure and pain in our lives. Sound perception was key to our survival until recent times when its importance was diminished with the advent of secure shelter, but it still affects each of us profoundly. Only recently through credible scientific studies have we been able to understand why sound affects us in so many different ways. Parallel to this scientific exploration of the hearing mechanism has been the development of the field of acoustics, which has provided methods for describing sound behavior and rating its associated intensities. Descriptors in any technical field can be confusing without proper training, and the field of acoustics provides ample material to feed that confusion.

Regulations have been introduced since the 1970s to address the potential negative health effects associated with noise exposure, but the extent of those effects and a clear link between noise and anything other than hearing loss have not been adequately defined. The explosion of unfiltered information available to the public over the past decade through the internet has led to even more confusion and we are at a point at which it is difficult, if not impossible, for a person without technical knowledge in this field to separate credible from speculative information. It is with this in mind that this book has been written.

Karl Kryter wrote three seminal books on the negative effects of sound on people, published between 1970 and 1994 – *The Effects of Noise on Man* (1970), *Physiological, Psychological, and Social Effects of Noise* (1984), and *The Handbook of Hearing and the Effects of Noise* (1994) – each one building on the next. These volumes were comprehensive and technical, and many changes have occurred since they were published, especially in terms of research results and the types of noise sources of concern to the public. This book is not meant to replace any of the valuable contributions Dr. Kryter has made to the field of psychoacoustics and, besides these works, there is no single book available that summarizes research efforts related to the effects of sound on people.

This book is for the non-technical student interested in understanding the state of current research in this field. More than 1,000 references were reviewed and close to 500 were included as those being the most credible, unique, and relevant to the latest research results. The descriptors commonly used in these publications are explained, along with common misinterpretations and misuses of the descriptors from experience and review of speculative publications.

Chapter 1 starts with an assumption of no background in acoustics by explaining the most basic acoustic parameters involved with sound generation and propagation both indoors and outdoors. Chapter 2 builds on this foundation by explaining the most common descriptors used in these studies. A key point with these descriptors is consistency, as any conclusion can be drawn from a study by choosing descriptors that support the desired conclusions. Without consistency in descriptors and their proper use, there is no credibility in reported results. Chapter 3 gives an overview of the hearing process, explaining generally how it works and what can happen when its delicate mechanism is not operating in perfect order. Alternate means of hearing (beside the normal channel) are described, along with an introduction to hypersensitivities that have not received much serious attention.

Chapter 4 summarizes the state of research in negative physiological effects associated with sound, from well-established results in noise-induced hearing loss to lesser-known ongoing research addressing the links between sound exposure and cardiovascular diseases, along with low-frequency and infrasound concerns. Chapter 5 summarizes the state of research in negative psychological effects associated with sound, covering the most-studied topics of annoyance, stress, sleep disturbance, learning disabilities, and emotional effects.

Chapter 6 continues with descriptions of the characteristics of current sound sources associated with negative sound effects to explain the aspects of these sources that contribute to their negative effects. Included in this discussion are transportation sources (roadway, aircraft, and rail), industrial sources (including traditional power plants and wind farms), recreational sources (such as firearms, public performances, toys, personal listening devices, and tools), hums (sounds only heard by some with no obvious origin), and the fallacies and realities of acoustic weapons.

Topics not often seen in these types of books are those related to the positive effects of sound on people. It is important to consider the positive as well as the negative effects when addressing the effects of sound to determine the most practical and effective alternatives to solving sound issues. In this regard, Chapter 7 summarizes the state of research in music psychology, sound therapies, soundscapes, and the ways in which sound is used to influence human behavior in common public environments. The book then finishes with the topics of sound control and regulation in Chapter 8, explaining noise control design and regulatory methods with common limits to inform the reader of the practical options available in dealing with negative sound issues. A glossary completes the book as a handy reference for explaining the many technical terms used in the book and public documents associated with this topic.

As mentioned above, more than 1,000 references were consulted for this book, and this would not have been possible without the invaluable services provided to me through the Boston Architectural College, where I've been teaching acoustics courses online for the past 16 years. My sincere appreciation goes out to the library staff under the leadership of Susan Lewis and Whitney Vitale, namely, Robert Adams, Erica Jensen, Toshika Suzuki, Sheri Rosenzweig, Rebecca Baker, Geoffrey Staysniak, Celia Contelmo, Christina Leshock, and Kris Liberman.

Jim Cowan
August 2015

1

Acoustic Parameters

1.1 Introduction

Acoustics is the science of sound. It involves many scientific disciplines, most notably physical, mechanical, electrical, biological, and psychological components. This interdisciplinary branch of science has permitted us to evaluate and control sound both to our advantage and to our detriment. Although hearing is not an essential element in acoustics, it has been the basis for our evaluations of sound over the centuries. As one of the most important mechanisms in our survival, the sense of hearing and the interpretation of sound shape our world.

Any discussion about the effects of sound on people must begin with an explanation of the parameters associated with sound generation, propagation, description, and perception. Without an understanding of these principles, a discussion about the effects of sound would not provide any meaningful information to the reader. This chapter covers sound generation and propagation, describing the most common ways in which a sound wave is altered as it travels from its source to a listener.

1.2 Sound Generation

Sound energy is generated when a medium is disturbed by particle motion. This disturbance generates pressure variations in the medium. These pressure variations travel in patterns associated with medium conditions and dissipate as they expand from a local source over an increasingly larger area. A simple two-dimensional representation of this can be visualized when a still body of water is disturbed by a small object or drop of water at a single location, as shown in Figure 1.1. The ripples in the water show peaks and valleys of pressure variations radiating out from the single point of contact.

The Effects of Sound on People, First Edition. James P. Cowan.
© 2016 John Wiley & Sons, Ltd. Published 2016 by John Wiley & Sons, Ltd.

Figure 1.1   Water disturbance pattern illustrating wave propagation from a point source in two dimensions

Sound energy in air radiates from a stationary source in a similar pressure pattern but in three dimensions. This pattern is characterized mainly by three parameters that are mathematically interdependent – frequency, wavelength, and wave speed. The main distinguishing factor between this type of energy and all others is that it can be detected by a hearing mechanism and interpreted for some form of action or communication. For the purposes of the information in this book, the term "sound" refers to any energy that is capable of stimulating the human hearing mechanism, as described in Chapter 3. The term "noise" refers to a subset of sound that is interpreted by humans as negatively affecting their environment. Sound therefore does not require personal interpretation, as noise is a subjective qualification. Sound exists in the forest if a tree falls and no one is there to hear it, but that same tree falling would not generate noise unless someone is there not only to hear it but also to interpret it as having a negative quality.

Sound requires a medium for the energy to propagate to a listener. It does not exist in a vacuum or in outer space. The big bang at the beginning of our universe generated no sound, although the word "bang" certainly implies the generation of sound. The key attribute of sound is that it its energy is of a form capable of stimulating a hearing mechanism.

1.2.1   Frequency

The simplest sound pressure pattern is generated by a source having pressure variations occurring at a constant rate, known as a pure tone. This would result in a sinusoidal pattern traveling away from the source as a wave, as shown in Figure 1.2, with the acoustic pressure oscillating with respect to equilibrium (the 0 position in Figure 1.2) at atmospheric pressure. This sinusoidal pressure pattern occurs with respect to both time and distance from the

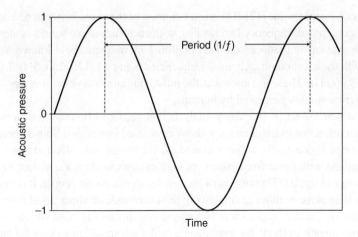

Figure 1.2 Acoustic pressure pattern for a single frequency (pure tone) with time

source. The time elapsed between repeating parts of the pressure pattern is known as the period of the wave, in units of cycles. The rate at which this pressure variation takes place is the reciprocal of the period, and is designated as the frequency, in units of cycles/second (s). The unit of cycles/s is most commonly denoted as Hertz (Hz), named for the German physicist Heinrich Hertz (1857–1894), who is primarily known for proving the existence of electromagnetic waves.

Humans can generally hear acoustic energy between 20 and 20,000 Hz but with varying sensitivity in that range. We are most sensitive to sounds in the 2,500–4,000 Hz range due to ear canal amplification (to be discussed further in Chapter 2), which is also the critical frequency range for speech intelligibility through consonant sound recognition. Although we can hear sounds below 20 Hz and above 20,000 Hz, their pressures must be many orders of magnitude higher than those in the 2,000 Hz range to be perceived at the same loudness.

Sounds with dominant energy below 20 Hz are categorized as infrasound and sounds with dominant energy above 20,000 Hz are labeled as ultrasound. There has been a significant amount of research and attention paid to infrasound and its potential effects on people and this is discussed in later sections of this book. Except at very high levels, infrasound is not audible and can be perceived as vibratory feelings in the body. There is conflicting information in the literature regarding the need for auditory perception for a sound to have any effect on people and this information is discussed in Chapter 4.

Ultrasound is not known to cause any noticeable effects on people but may affect other species due to variations in frequency sensitivities between species. For example, bats and dolphins rely on ultrasound to navigate and communicate, with peak sensitivities in the 20,000–80,000 Hz range [1].

As this book is focused on the human experience, only sound energy dominated in frequencies below 20,000 Hz is discussed. For those with musical knowledge, middle C on the piano keyboard is roughly 262 Hz. The American National Standards Institute has standardized the use of specific preferred frequencies to evaluate sound energy, based on $10^{0.1N}$

(where N is an integer value) [2]. The most common of these are between 63 and 8,000 Hz in constant-percentage frequency bandwidths, with each successive band's center frequency being twice that of its predecessor. This doubling of frequencies is known as an octave increase, with the most commonly used frequencies being 63, 125, 250, 500, 1,000, 2,000, 4,000, and 8,000 Hz. These are known as the most common octave band center frequencies used to describe sounds perceived by humans.

One important aspect of frequency analysis that seems to be confused in much of the literature is a reference to frequencies without associated intensities. Sound energy in every frequency range is constantly varying around us, but we are only affected when the energy levels associated with those frequencies are high enough to elicit a reaction. For example, if sound energy in the 100 Hz range is of interest for its effects on people, it is inappropriate to consider it an issue without also knowing the magnitude or sound level associated with energy in that frequency range. There is a magnitude threshold below which sound energy will not cause negative effects for most people (although sensitivities vary for each person) and that level varies with frequency.

Pure tones (with dominant acoustic energy at a single frequency) rarely exist in nature and are sometimes generated by man-made sources. For the most part, however, the sounds we are exposed to are composed of contributions from energy at all audible frequencies.

Musical sounds and some sounds generated by machinery are often composed of energy peaks at integer multiples of a base frequency, called the fundamental frequency. The integer multiples of the fundamental frequency are often called harmonics. Harmonics tend to enhance the enjoyment of sounds in music but that is not always the case for other types of sources. Examples of acoustic signatures incorporating harmonics are mentioned in later sections of this book.

1.2.2   Wavelength

As mentioned earlier, the sinusoidal pressure wave pattern for a pure tone occurs in terms of both time and distance. When viewing this variation in terms of distance, as shown in Figure 1.3, the distance between repeating parts of the wave is known as the wavelength. Wavelength is associated with frequency according to the following equation, as long as the speed of the sound wave is constant:

$$\lambda = c/f \tag{1.1}$$

where c is the speed of sound, f is the frequency, and λ is the wavelength.

This demonstrates an inverse relationship between frequency and wavelength, which is revealed in Table 1.1 (based on equation 1.1).

Considering that materials effective in controlling sound must at a minimum have dimensions that are comparable to a wavelength of the sound of interest, Table 1.1 shows that higher frequencies are much easier to control than lower frequencies, given the size differential. For example, material dimensions in excess of 4 m would be required to affect a sound source with dominant energy below 100 Hz, while material dimensions less than 4 cm would effectively control a sound source with dominant energy higher than 8,000 Hz.

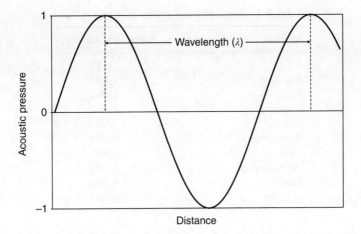

Figure 1.3 Acoustic pressure pattern for a single frequency (pure tone) with distance

Table 1.1 Correlation between frequency and wavelength in air at 20°C

Frequency (Hz)	Wavelength (m)
20	17
100	3.43
343	1
1,000	0.343
4,000	0.086
8,000	0.043
16,000	0.021

The speed of sound depends on several factors, including temperature and the density of the medium through which it is travelling. Therefore, any changes in temperature or density will cause the speed of sound to change. Table 1.2 lists the speed of sound for various media at 20°C. The variation of the speed of sound with temperature in air can be determined from equation (1.2):

$$c = 331.3 + 0.61T \qquad (1.2)$$

where c is the speed of sound in m/s and T is the temperature in °C.

1.3 Sound Propagation

As a sound wave travels away from its source, it encounters many conditions that affect its characteristics when it arrives at a listener. If there are no obstructions in its path and no change in medium conditions, sound intensity will decrease with distance from a source at a rate associated with the surface area of the expanding wavefront (perpendicular to the direction of travel).

Table 1.2   Speed of sound in various media at 20°C

Medium	Speed of sound (m/s)
Air	343
Sea water	1,500
Plexiglas	1,800
Concrete	3,400
Copper	3,500
Wood/marble	3,800
Steel	5,050
Aluminum	5,140
Glass	5,200
Gypsum	6,780

This is known as divergence. In addition to divergence, sound waves are affected by atmospheric, topographic, and ground conditions outdoors, and enclosure shapes and materials indoors.

1.3.1   Unimpeded Divergence

A stationary sound source that is small in size compared with the propagation distances being considered is known as a point source. Unimpeded sound energy radiating from a point source propagates in a spherical pattern, as shown in Figure 1.4. Rather than the distance-varying pressure wave shown in Figure 1.3, Figure 1.4 shows the propagation of

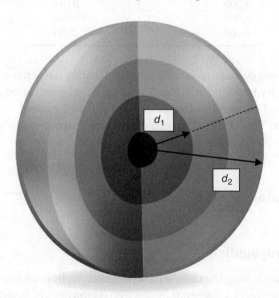

Figure 1.4   Acoustic pressure wave pattern showing wavefront propagation from a point source in three dimensions

sound by wavefronts perpendicular to the direction of travel. These can be thought of as the three-dimensional pressure peaks propagating from a single disturbance as illustrated with water waves in Figure 1.1. The energy associated with a source is a constant value, but the energy at specific locations distant from a source will dissipate, as the total sound energy is spread over an increasingly larger area. As the surface area of a sphere is $4\pi d^2$, where d is the distance from the center of the sphere, the acoustic energy drops off at a rate proportional to the square of the distance from a point source.

A series of moving point sources (such as a steady stream of vehicular traffic on a roadway) or a sound source resembling a continuous line more than a point (such as a long train or electrical transmission line) is known as a line source. Unimpeded sound energy radiating from a line source propagates in a cylindrical pattern, as shown in Figure 1.5. As for a point source, energy at specific locations distant from a line source will dissipate, as the total sound energy is spread over an increasingly larger area. As the surface area of a cylinder is $2\pi d$, where d is the distance from the center of the cylinder, the acoustic energy drops off at a rate proportional to the distance from a line source.

1.3.2   Impeded Propagation

The properties of sound waves change when they interact with variations in conditions in their travels between a source and listener. Sound energy is affected by changes in media as well as changes in properties within a medium. There are generally four types of phenomena that result from these changes encountered by sound waves – reflection, refraction, diffraction, and diffusion.

All of these properties are analogous to the laws of optics regarding the behavior of light when it encounters changes in media. Although sound and light are based on different types of energy (sound is based on mechanical and light is based on electromagnetic energy),

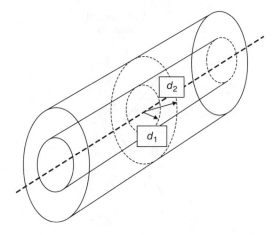

Figure 1.5   Acoustic pressure wave pattern showing wavefront propagation from a line source in three dimensions

they are both based on wave motion exciting sensations in human receptors. They are both described in terms of frequency and wavelength, yet they are in vastly different ranges. Specifically, the smallest wavelength that can typically be heard by humans is on the order of 20,000 times larger than the largest wavelength that can be seen.

Reflection

When a sound wave encounters a sharp discontinuity in medium density, a portion of its energy is reflected at the interface between the medium changes. If that interface is a hard, smooth surface, the angle of incidence of the wavefront (Θ_i) equals the angle of reflection (Θ_r), as shown in Figure 1.6. The portion of the sound energy not reflected at the interface is either transmitted through the interface or dissipated as heat energy at the interface.

Reflection also occurs, within the same medium, when a sound wave encounters a sharp discontinuity in cross-sectional area along its propagation path. This is caused by a mismatch in acoustic impedance, which is a function of the cross-sectional area of the propagation path.

Refraction

Acoustic refraction is analogous to the refraction of light, which is governed by Snell's law, named for the Dutch astronomer and mathematician Willebrord Snellius (1580–1626). Similar to reflection, refraction occurs when a sound wave encounters a change in medium conditions. Unlike reflection, refraction changes the direction of sound propagation into the adjacent new medium condition rather than causing it to direct sound energy back into the incident sound wave's medium. It is similar to what happens to light as it shines through

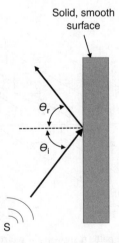

Figure 1.6   Reflection of an acoustic wave off a smooth, solid surface

a body of water or travels through a prism. This change in direction of sound travel results from a change in the speed of sound. This phenomenon is summarized in equation (1.3).

$$c_2/c_1 = \sin\theta_r/\sin\theta_i, \qquad (1.3)$$

where c_2 is the speed of sound in medium 2, c_1 is the speed of sound in medium 1, θ_r is the angle of the refracted wave travel with respect to the perpendicular to the medium interface, and θ_i is the angle of the incident wave travel with respect to the perpendicular with the medium interface.

This relationship is illustrated in Figure 1.7 for the conditions of increasing and decreasing speeds of sound. Snell's law basically states that the direction of sound travel will change as the medium or medium conditions change. As the speed of sound varies with temperature, as described by equation (1.2), sound waves traveling outdoors through areas with varying temperatures will change propagation direction accordingly. This phenomenon is described in more detail later in this chapter.

Diffraction

Diffraction occurs when a sound wave encounters a solid barrier or an opening in a barrier. As for light being blocked by a barrier (but not a complete enclosure), some sound energy will be reduced but most will bend over and around the barrier. There is a limit to the sound

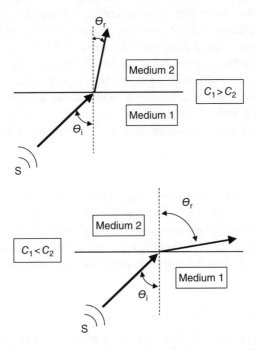

Figure 1.7 Refraction of an acoustic wave through changes in medium conditions

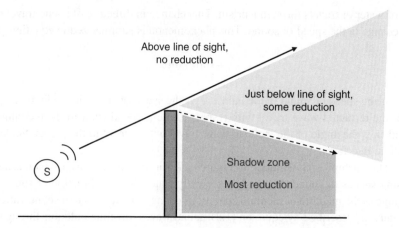

Figure 1.8   Diffraction of an acoustic wave over a solid barrier

reduction capability of a barrier because of diffraction, and that limit is independent of the barrier composition. Figure 1.8 shows a simplification of the noise reduction effectiveness of a typical barrier.

Using the light analogy, the region with the greatest reduction is known as the shadow zone. Bear in mind that diffraction effects only occur within 100 m of a barrier, assuming the barrier is less than 30 m from the sound source of interest, and the largest effects are close to the barrier. Beyond these distances, diffraction effects are minimal. The key parameter for sound reduction due to diffraction is breaking the line of sight between the source and listener. No diffraction-related sound reduction occurs above the line of sight. Therefore, any portions of residences visible by vehicles driving on a roadway lined with noise barriers would not receive any acoustical benefit from that barrier.

Diffraction is frequency-dependent. For significant diffraction to occur, the physical dimensions of a barrier must be at least comparable to the wavelength of the lowest frequency of interest. As lower frequencies are associated with larger wavelengths, sound reduction from diffraction is typically limited to frequencies above 250 Hz.

Diffusion

Diffusion is the even distribution of sound energy after a sound wave reflects off an uneven or convex reflective surface. Diffusion is used in architectural acoustics to eliminate sharp echoes without sacrificing reflected sound energy to provide an even spreading of sound throughout an audience area. Because it is frequency-dependent, diffusion depends on the size of the materials, so, as for diffraction, it is usually practical for frequencies above 250 Hz only.

1.3.3   Sound Behavior Indoors

As many sound sources that affect people are in enclosed spaces, it is useful to understand some general ways in which the indoor environment uniquely affects sound waves.

The shapes and finish of materials in rooms can have a significant effect on sound energy. The most prevalent acoustic phenomena associated with reflective surfaces in enclosed spaces are echoes, room modes, and reverberation.

Echo

Typically, if the arrival times between two distinct sound waves at the human ear is less than 50–100 milliseconds (ms), the two sounds are blended and perceived as a single sound wave by the brain. When that arrival time is greater than 100 ms, the two sounds are heard separately, creating the perception of an echo. This phenomenon is discussed further in Chapter 3. Figure 1.9 shows an example of how an echo can be generated. Considering the speed of sound in air at 20°C of 343 m/s, a sound wave would travel a distance of roughly 34 m in 100 ms. Therefore, if the difference between the direct (D) and total length of the reflected ($R_1 + R_2$) path in Figure 1.9 exceeds 34 m, the reflected sound wave would be heard separately from the direct sound wave and an echo would be interpreted.

Room Modes

Room modes, also known as room resonance, occur between parallel reflective surfaces in two and three dimensions. When the distance between these surfaces is equal to an integer multiple of half-wavelengths of sound, the incident and reflected pressure patterns combine to generate a pressure pattern that stands still in the room. An example of one of these so-called standing waves is shown in Figure 1.10. These modes are characterized by having locations within a room at which there are regions of minimum acoustic pressure at nodes

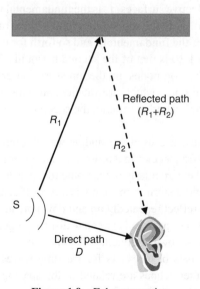

Figure 1.9   Echo generation

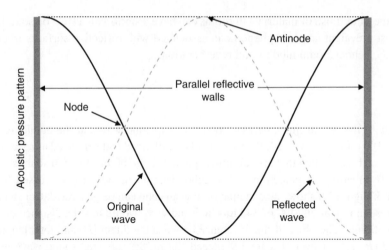

Figure 1.10   Standing wave generation from parallel reflective surfaces

where the incident and reflected waves combine destructively to cancel the amount of total pressure and maximum acoustic pressure at antinodes where the incident and reflected waves combine constructively to amplify the total pressure.

The frequencies at which these standing waves occur in two dimensions are calculated as follows:

$$f_n = nc/2D \tag{1.4}$$

where n is a positive integer, f is the frequency, c is the speed of sound, and D is the distance between parallel reflective surfaces. f_1 is the fundamental frequency (corresponding with half of a wavelength), f_2 is the second harmonic (twice the fundamental), f_3 is the third harmonic (three times the fundamental), and so forth for increasing values of n. The pressure pattern in Figure 1.10 is that of the second harmonic. The harmonic number, n, corresponds to the number of nodes in the pressure pattern. These are sometimes described in terms of overtones, which are integer multiples of the fundamental, such that the first overtone is the second harmonic, the second overtone is the third harmonic, and so forth.

The variable pressure patterns caused by standing waves combined with the property that they only occur at specific frequencies in a room can cause significant distortions in indoor sound patterns. These can be particularly bothersome for speech communication in rooms.

Another cause of acoustic distortion in rooms is a type of echo known as flutter, which occurs when sound waves reflect repeatedly up and down or side to side along tall parallel reflective surfaces. For narrow spacings between walls, flutter generates buzzing or ringing sounds. For wider spacings, lower-frequency flapping sounds are produced, similar to those generated by flying bats or birds. As for standing waves, the dominant frequency ranges associated with flutter echoes are related to the spacing between parallel walls – wider spacings are associated with lower frequencies.

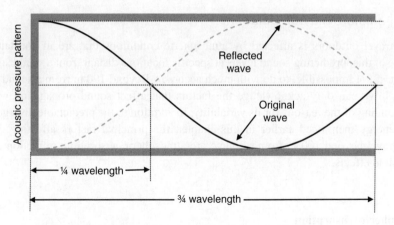

Figure 1.11   Standing wave generation from a closed–open system

A different type of room mode is generated when one end of a room is open, as shown in Figure 1.11. The acoustic impedance (the ability of a system or condition to impede the flow of sound energy) is related to the volume of space through which a sound wave travels. As mentioned earlier, a significant discontinuity in cross-sectional area in the sound path introduces an impedance mismatch causing sound energy to reflect back from the open end of a closed-open system.

The frequencies at which these standing waves occur in two dimensions are:

$$f_n = (2n-1)c/4D \tag{1.5}$$

where n is a positive integer, f is the frequency, c is the speed of sound, and D is the distance from the closed end to the open end of the system. The fundamental frequency for this system corresponds to one quarter of a wavelength. The pressure pattern shown in Figure 1.11 is for the second harmonic and, as for the room with parallel reflective walls, the harmonic number (n) is the number of nodes in the pressure pattern. This system provides the physical mechanism by which sounds are generated from organ pipes, and brass and woodwind instruments.

Reverberation

Reflections from non-parallel surfaces in an enclosure create a blending of sound waves known as reverberation. Contrasted with room resonance, reverberant fields in rooms are characterized by areas of relatively constant sound level due to the random interaction of reflected energy throughout the space. Reverberation is mainly dependent on the open volume of a space and the reflective qualities of its finishes, and it can significantly affect speech intelligibility. It is typically described quantitatively by reverberation time (RT_{60}), which is explained further in Chapter 2.

1.3.4   Sound Behavior Outdoors

Sound travel outdoors is affected by atmospheric conditions that are in constant flux. Because of this, predicting sound levels at specific locations distant from a sound source is difficult, if not impossible, to do with much accuracy. Beyond 100 m from a sound source outdoors, we can only acknowledge the factors that affect sound propagation with the understanding of the reasons for its variability. In addition to the predictable divergence of sound energy mentioned earlier in this chapter, the principal factors affecting outdoor sound travel beyond 100 m from a source are atmospheric absorption, refraction effects, and ground effects.

Atmospheric Absorption

Atmospheric absorption takes place at the molecular level and is highly frequency-dependent. The conditions that affect atmospheric sound absorption the most are temperature and humidity. Figure 1.12 shows the extent of these effects with temperature and Figure 1.13 shows the effects with humidity. In each case, these effects are minimal below 500 Hz and increase dramatically with frequencies above 2,000 Hz. The decibel (dB) scales shown in these figures are explained further in Chapter 2.

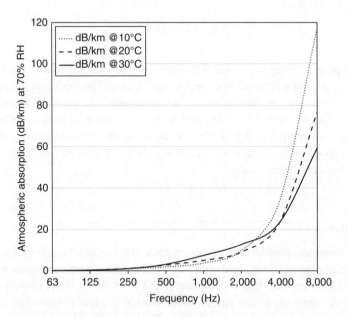

Figure 1.12 Atmospheric absorption factors with temperature. RH, relative humidity. (Adapted from data in ISO 9613-2:1996 [3].)

Atmospheric Refraction

The greatest variation in outdoor sound levels distant from a source occurs due to atmospheric refraction. According to equation (1.3) and as illustrated in Figure 1.7, Snell's law dictates that the direction of sound wave propagation will change if the speed of sound changes within a medium. The two atmospheric conditions that cause the most dramatic refraction effects are changes in temperature and wind currents.

According to equation (1.2), the speed of sound in air increases with increasing temperature. This translates to sound waves bending toward cooler temperatures independent of all other atmospheric conditions. Figure 1.14 illustrates typical atmospheric conditions on a clear afternoon when temperatures decrease with increasing elevation, also known as temperature lapse conditions. In this case, sound waves would refract upward toward cooler temperatures, creating a shadow zone (using the optics analogy of reduced light)

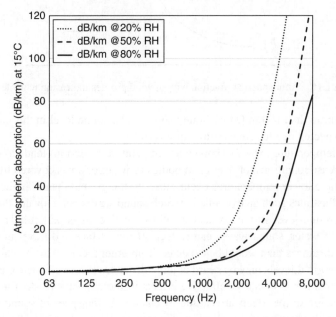

Figure 1.13 Atmospheric absorption factors with humidity RH, relative humidity. (Adapted from data in ISO 9613-2:1996 [3].)

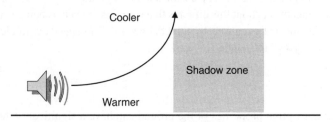

Figure 1.14 Atmospheric refraction with temperature lapse conditions

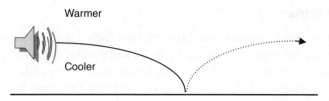

Figure 1.15   Atmospheric refraction with temperature inversion conditions above a reflective surface

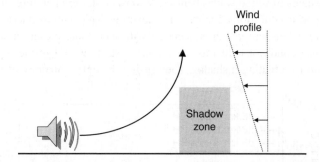

Figure 1.16   Atmospheric refraction with upwind profiles increasing with elevation

for listeners more than 100 m from a sound source. The sound level in this case would be lower than expected due to temperature lapse refraction.

The other temperature extreme (known as temperature inversion) tends to occur late at night, in overcast conditions, or over calm bodies of water or areas covered by frozen precipitation, where lower temperatures occur closer to the ground than at higher elevations. Figure 1.15 illustrates such an example in which sound waves bend toward the cooler surface. If the ground-level surface is acoustically reflective, as is particularly the case for calm bodies of water, sound waves can reflect off the surface and travel unimpeded for much longer distances than would be expected from other means. This explains why conversations at normal levels may be heard at large distances across opposite ends of a lake.

Wind currents can also cause refractive effects in the atmosphere. As for temperature changes, these refractive effects are caused by changes in the speed of sound as wind currents change with elevation, as are illustrated in Figures 1.16 and 1.17. As the figures show, upwind variations create shadow zones and downwind variations cause sound waves to travel with less attenuation than expected for distances greater than 100 m from a source. A common misconception about this effect is that sound is viewed as being carried with the wind; however, this effect is caused by the changes in wind speeds with elevation rather than the wind "carrying" the sound wave.

Ground Effects

The acoustical absorption properties of the ground can have a significant effect on attenuating sound propagation, especially for a source and listener within 10 m of an absorptive surface. This effect is frequency-dependent, and is shown for an absorptive surface in Figure 1.18. This effect is most pronounced in the 250 Hz range and is negligible above

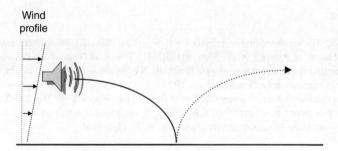

Figure 1.17   Atmospheric refraction with downwind profiles increasing with elevation

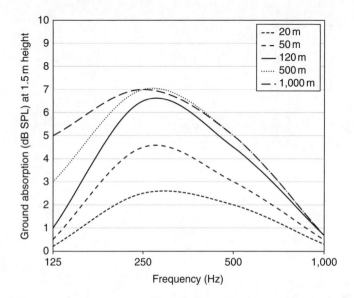

Figure 1.18   Ground absorption factors with distance from a source over an absorptive surface. SPL, sound pressure level. (From data in ISO 9613-2:1996 [3].)

1,000 Hz. Ground surfaces that have the highest acoustical absorption properties are porous, e.g., loose soil, fresh snow, and dense gravel. Rock, pavement, and calm bodies of water are the most reflective (or least absorptive) surfaces, which can amplify sounds at distances from a combination of reflection and refraction.

Vegetation Effects

Vegetation provides little, if any, extra sound attenuation unless it is at least 30 m in depth and provides complete visual occlusion of the source to the listener. The greatest potential effect of dense vegetation is enhancing the ground absorption effect [4]. Although often used for noise reduction purposes, a single line of trees or bushes between a source and listener provides no measurable decrease in sound to the listener, even if the vegetation blocks the line of sight to the source.

References

[1] Stebbins, W.C. *The Acoustic Sense of Animals*. Cambridge, MA: Harvard University Press, 1983.

[2] Acoustical Society of America. ANSI S1.6-1984 (R2011). *Preferred Frequencies, Frequency Levels, and Band Numbers for Acoustical Measurements*. New York, NY: American Institute of Physics, 2011.

[3] International Organization for Standardization. ISO 9613-2:1996. *Acoustics – Attenuation of Sound during Propagation Outdoors – Part 2: General Method of Calculation*. Geneva, Switzerland: ISO, 1996.

[4] Price, M.A., Attenborough, K., and Heap, N.W. (1988). "Sound attenuation through trees: Measurements and models." *Journal of the Acoustical Society of America*, 84(5): 1836–1844.

2

Sound Description

2.1 Introduction

The next step in laying the foundation for an understanding of the effects of sound on people is an explanation of the descriptors most commonly used to quantify sound energy. There are many methods used for describing sound and all of them can be confusing to anyone unfamiliar with them. This chapter covers the most common sound descriptors found in standards, guidelines, and research reports.

2.2 The Decibel Scale

The human hearing mechanism is sensitive to a large range of pressure variations, with the threshold of pain associated with a pressure 1,000,000 times greater than that for the threshold of hearing. Combining this vast pressure range with the conditions for which a doubling of sound energy is just noticeable to most people and a 10-fold increase doubles the apparent loudness, the logarithmic scale is much more appropriate for describing sound than a linear scale.

A logarithm is an exponent of a stated base value. The scale adopted for describing sound energy uses 10 as the logarithmic base. For example, the logarithm to the base 10 of 100 is 2, as 100 is 10^2, and the logarithm to the base 10 of 1,000 is 3, as 1,000 is 10^3. Since 10^0 is 1, the logarithm to the base 10 of 1 is 0. Some mathematical properties of logarithms to keep in mind are as follows:

$$\log\left(a^n\right) = n\log\left(a\right) \tag{2.1a}$$

The Effects of Sound on People, First Edition. James P. Cowan.
© 2016 John Wiley & Sons, Ltd. Published 2016 by John Wiley & Sons, Ltd.

$$\log(ab) = \log(a) + \log(b) \tag{2.1b}$$

$$\log(a/b) = \log(a) - \log(b) \tag{2.1c}$$

The term "sound level" implies a logarithmic ratio of parameters associated with sound energy, denoted in units of bels, named for the Scottish inventor Alexander Graham Bell (1847–1922). As the base unit of a bel is too coarse for rating human sound perception, one-tenth of a bel, or a decibel (dB), was adopted as the standard unit for rating sound energy. The basic definition of a sound level rating is then:

$$L_E = 10\log(E/E_{ref}) \tag{2.2}$$

where L_E is the acoustic energy level in decibels (dB), E is the actual acoustic energy, and E_{ref} is a reference energy.

Acoustic pressure is the parameter that is sensed by the human hearing mechanism as well as measured by microphones, so this is the quantity of most interest to anyone evaluating the effects of sound on people. As energy is proportional to the square of the pressure, equation (2.2) can be defined in terms of the sound pressure level (denoted as SPL or L_p) as:

$$SPL = 10\log(p^2/p_{ref}^2) \tag{2.3}$$

where p is the actual pressure and p_{ref} is the reference pressure.

Using the property of logarithms in equation (2.1a), equation (2.3) can be rewritten as:

$$SPL = 20\log(p/p_{ref}) \tag{2.4}$$

which is the most widely used definition of SPL, where p_{ref} is the pressure associated with the average threshold of hearing for a person with a healthy hearing mechanism, or 20 micropascals (μPa, where 1 μPa $= 1 \times 10^{-6}$ Pa). The pressure unit of pascals (named for the French mathematician and physicist Blaise Pascal [1623–1662]) is also represented by the units of newtons/square meter (N/m²; named for the English mathematician and physicist Isaac Newton [1643–1727]), pressure being force per unit area. p_{ref} was chosen as the threshold of hearing to establish an SPL of 0 dB as the threshold of hearing, or the SPL below which most people cannot hear. As mentioned earlier, the logarithm of 1 is 0 and the ratio of p/p_{ref} is 1 when $p = p_{ref}$. A negative SPL implies that the pressure is not audible to the average listener.

As acoustic pressure varies with location and distance from a source, a distance reference must be included with an SPL specification to give it any meaning. Sound power level, on the other hand, is characteristic of a source and is independent of location. Sound power is proportional to acoustic energy and sound power level (denoted as PWL, or L_w), is defined mathematically as:

$$PWL = 10\log(W/W_{ref}) \tag{2.5}$$

where W is the power associated with the source (in watts [W]) and W_{ref} is 1×10^{-12} W, the power associated with the threshold of hearing. Sound power levels are often used in specifications for mechanical equipment. They cannot be directly measured, as SPLs can be. They can only be calculated based on many SPL measurements close to and surrounding a source.

The relationship between SPL and PWL is defined by:

$$PWL = SPL - 20\log(d) - 11 \tag{2.6}$$

where d is the distance between the source and reference SPL location (in m). This assumes a sound source in free space not close to any large reflective surfaces. If the source is near one or more large reflective surfaces, the term, $10\log(2^n)$ should be added to equation (2.6), where n is the number of large reflective surfaces nearby. Therefore, if the source is near a large reflective floor or ground surface, 3 dB would be added (10 log2) and if the source is in the corner of a room (where three walls come together), 9 dB would be added (10 log8). This is the directivity term, which adds SPL at a distance from a source by confining the propagation area from what would occur in free space. The basic point source divergence assumption of spherical spreading is confined to a specific direction from nearby reflective surfaces in this case, making a sound source in the corner of a room potentially louder to anyone in the room than if the source were placed in the center of the room. This is the principle behind placing horns on loudspeakers to focus most of the sound energy in a specific direction.

2.3   Frequency Weighting Networks

As introduced in Chapter 1, the human hearing frequency range is generally regarding as 20–20,000 Hz, but our sensitivities vary for sounds within that frequency range. One reason for this is ear canal resonance, which amplifies sounds in the 2,500–4,000 Hz range for adults (and at up to twice that frequency range for young children). The human ear canal is essentially a cylinder open to the air at one end and closed off by the eardrum at the other, making it susceptible to the closed–open system resonance phenomena illustrated in Figure 1.11. As the average adult human ear canal is 2.3–2.97 cm in length [1], the fundamental resonance frequency of this system is 2,887–3,728 Hz at 20°C, using equation (1.5). Figure 2.1 shows the resultant amplification in human hearing in this frequency range, based on the average ratio of sound pressure at the eardrum to that at the entrance of the ear canal, rather than the actual processed sounds through the entire hearing mechanism. The operation of the entire hearing mechanism is discussed in Chapter 3.

2.3.1   Loudness

The threshold of human hearing is well documented and illustrated in Figure 2.2 over the 20–20,000 Hz range. As this shows, sounds are discounted in all frequency ranges except for the ear canal resonance range of 2,500–4,000 Hz. Although there is an amplification in

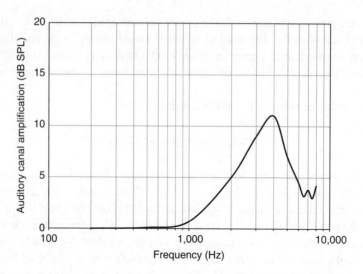

Figure 2.1 Human ear canal amplification. SPL, sound pressure level. (From Wiener [2] with permission of the Acoustical Society of America)

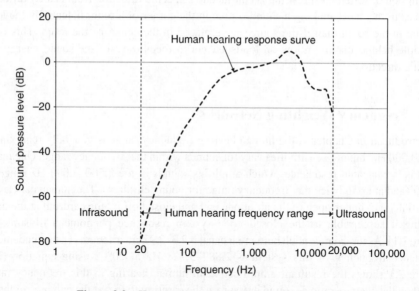

Figure 2.2 Human threshold of hearing with frequency

hearing sensitivity in the ear canal resonance frequency range, it is not as pronounced as is shown in Figure 2.1 from the resonance of the ear canal itself due to signal processing between the eardrum and the brain. There is a sharp drop-off in sensitivity below 500 Hz and above 8,000 Hz. This means that the level of a signal with most of its energy below 500 Hz or above 8,000 Hz would have to be much higher than one with most of its energy between 500 and 8,000 Hz to be perceived as having the same loudness; however, this sensitivity pattern varies with sound level.

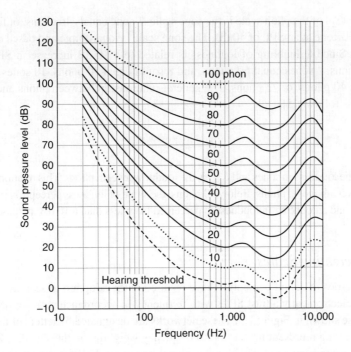

Figure 2.3 Human hearing equal-loudness curves (ISO 226:2003; [4] permission to reproduce extracts from ISO 226:2003 is granted by BSI. British Standards can be obtained in PDF or hard copy formats from the BSI online shop: www.bsigroup.com/Shop or by contacting BSI Customer Services for hardcopies only: Tel: +44 (0)20 8996 9001, Email: cservices@bsigroup.com)

This point is illustrated in Figure 2.3. Decades of measurements [3] have resulted in a set of so-called equal-loudness curves, standardized by the International Organization for Standardization (ISO) [4]. These curves represent the same loudness judged by a person with normal hearing, aged 18–25, for tones at different frequencies based on a reference SPL at 1,000 Hz. The unit prescribed for these curves is a phon. Note that the SPL at 1,000 Hz for each of these curves is the same as the phon rating for each of the curves. 1,000 Hz was chosen as the reference frequency for the phon scale because there is no amplification or reduction of sound at that frequency by the human hearing mechanism (shown as 0 on the frequency response curve).

The phon scale has been used to rate the loudness of household appliances. Since the logarithmic scale already used for decibel description can be confusing to the public, the phon scale can add to that confusion and is best used to compare the loudness of different sources rather than to rate the loudness of an individual source. The 10 and 100 phon curves are dotted in Figure 2.3 because they represent a limited amount of supporting data and are therefore extrapolated and not as supported by research as the other curves.

Beside the ear canal resonance and reduction of sensitivity to sounds below 500 Hz and above 8,000 Hz, it should be noted that the slope of the sensitivity reduction changes with increasing sound level. The 10 phon curve shows a much sharper rate of reduction in sensitivity for frequencies below 500 Hz than the 70 phon and higher curves.

One variation of the phon scale is the sone scale. 1 sone equals 40 phons, or the loudness of a 1,000 Hz tone at an SPL of 40 dB. The sone scale is logarithmic and based on loudness perception. Since a doubling of loudness is related to a 10 dB increase in SPL, 2 sones equal 50 phons, 4 sones equal 60 phons, and 8 sones equal 70 phons. 10 sones is 10 times louder than 40 phons, or 73 phons. The general relationship between phons and sones can be written as follows [5].

$$N = 2^{(L_{phon}-40/10)} \tag{2.7}$$

where N is the loudness in sones and L_{phon} is the loudness level in phons. This relationship is only valid, however, above 40 phons. Below 40 phons, the slope of loudness perception is not a linear relationship and a doubling in loudness is perceived from less than a 10 dB increase [6].

2.3.2   Weighting Scales

These properties of sound perception have resulted in the adoption of several "weighting" scales that electronically adjust SPLs with frequency in accordance with sensitivity scales such as those shown in Figure 2.3. These networks are designated by letters of the alphabet, with Z-weighting representing zero or no frequency weighting in the signal. A Z-weighted frequency response curve would be a horizontal line across the 0 relative response rating, as the signal is not electronically filtered.

As Figure 2.3 shows, human frequency sensitivity varies with sound level. The A-weighting network represents the frequency response of the 20–60 phon curves (but is actually based on the 40 phon curve), the B-weighting network represents the frequency response of the 70–90 phon curves, and the C-weighting network represents the frequency response above 90 phons. Most SPL ratings use the dBA scale, justifying that decision by stating that it matches human hearing frequency sensitivity. It is worth noting, however, that the dBA scale is based on moderate sound levels and the dBB or dBC scales are more appropriate for modeling human hearing response at elevated levels.

These scales have been standardized [7] and are illustrated in Figure 2.4. In each case, sound levels filtered using these scales are represented by dB followed by the network designation (e.g., dBA for A-weighted levels). The B-weighted scale is not used for most applications, but dBA and dBC values are commonly used. All of these scales have no weighting for signals at 1,000 Hz, which makes that the most practical frequency for calibrating instruments that use them.

Because dBA values discount low-frequency sounds to a greater extent than do dBC values, a comparison between dBA and dBC values for a source reading can offer a general qualification of the amount of low-frequency energy in a signal. dBA values are used most often for rating environmental noise and potential negative effects on people. Table 2.1 shows common sound sources and their associated dBA values in order to give perspective to the numbers.

dBC values will be higher than dBA values in most cases; however, due to the amplification of the dBA scale in the 1,250–5,000 Hz range (to account for ear canal

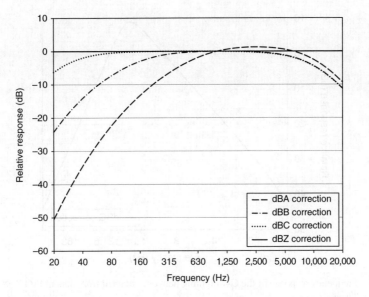

Figure 2.4 Frequency responses of the A-, B-, C-, and Z-weighting networks (adapted from data in ANSI S1.42-2001(R2011) [7])

Table 2.1 Sound pressure levels with associated common sources

Sound source/environment	Sound pressure level (dBA)
Siren at 15m/threshold of pain	120
Average levels at rock concerts	110
Average levels at clubs with music	100
On sidewalk by heavy truck or bus	90
On sidewalk by busy highway	80
On sidewalk by busy local street	70
Urban background	60
Suburban background	50
Quiet suburban area at night	40
Quiet rural area at night	30
Isolated broadcast studio	20
Audiometric booth	10
Threshold of hearing	0

resonance), it is possible for dBA values to exceed dBC values if the dominant sound energy is in the 1,250–5,000 Hz range. Even considering that aspect, dBA values cannot be more than 2 units higher than dBC values under any circumstances.

It is generally accepted that most people can just about notice a change of 3 dBA in SPL, a change of 5 dBA is clearly noticeable, and a change of 10 dBA represents a change in loudness by a factor of 2. For example, a 70 dBA signal would sound twice as loud as a 60 dBA signal, and an 80 dBA signal would sound four times as loud as a 60 dBA signal, while a 50 dBA signal would sound half as loud as a 60 dBA signal.

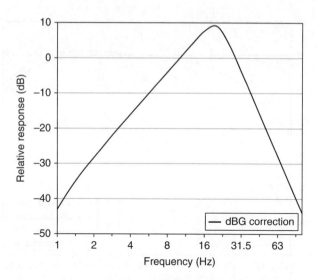

Figure 2.5   Frequency response of the G-weighting network (adapted from data in ISO 7196:1995 [8])

Another weighting network of common use is the G-scale, used for rating infrasound. This curve is shown in Figure 2.5, standardized by the ISO [8].The dBG scale has a zero weighting at 10 Hz and the network has a peak sensitivity to sounds in the 16–20 Hz frequency range, based on the potential annoyance of infrasonic signals to humans.

Other weighting networks have been developed over the past 50 years, such as dBD for aircraft noise, dBE for an alternative to dBA for more accurate perceived loudness rating (with E standing for ear [9]), and dBU for audible ultrasound, but none of these is commonly used.

These levels represent a single number rating for the total energy of an acoustic wave, including levels at all frequencies. When more detail is required to describe the frequency signature of a source, as would be appropriate for diagnosing noise problems with the intent of providing an appropriate solution, frequency analysis must be performed on the signal to ensure that measures appropriate for the frequency range of interest are used to control the sound.

2.4   Frequency Band Analysis

A spectrum is a plot of sound level on the vertical axis against frequency on the horizontal axis. As mentioned in Chapter 1, sound spectra are typically characterized by levels associated with constant-percentage frequency bands called octave bands (named from the eight – oct – standard notes on the musical scale), which are labeled according to their center frequencies, established in ANSI S1.6-1984 (R2011). Constant-percentage bands are evenly spaced when displayed on a logarithmic scale but not on a linear scale, with each successive bandwidth (frequency spacing) being a multiple of that of its predecessor. These frequency bands are characterized by an upper frequency limit (f_u), a lower frequency limit (f_l) which is half of the upper limit, and a center frequency (f_c), with the following relationship between them:

$$f_l f_u = f_c^2 \qquad (2.8)$$

The upper and lower limits of each octave band occur where the frequency response of the octave band filter drops off by a factor of 3 dB. The bandwidth is defined as the difference between f_u and f_l. The geometric center frequency of each band is the labeled frequency for each band, with the most common center frequencies being 63, 125, 250, 500, 1,000, 2,000, 4,000, and 8,000 Hz. Each subsequent center frequency is twice the value of its predecessor. Figure 2.6 shows a frequency response curve for a generic constant-percentage band and its adjacent bands. For each octave band frequency response to appear (on a chart) symmetrical and equal in bandwidth across several octaves, as is shown in Figure 2.6, the horizontal scale would need to be logarithmic as the bandwidth of each successive octave band is twice that of its predecessor.

When octave band analysis does not provide enough detail, finer bandwidths can be used, but 1/3-octave bands are typically the choice if slightly finer detail is required. These divide each octave band into three bands on a constant-percentage basis, with f_u being $2^{1/3}$ (or 1.26) multiplied by f_l.

Sound levels associated with different bandwidths cannot be simply combined, so it is best to combine or compare data having the same bandwidths. Otherwise, correction factors must be applied to account for the differing amount of acoustic energy within differing frequency bandwidths. The only condition under which this is not the case is for pure tones as, in those cases, most of the energy is confined in a bandwidth smaller than the

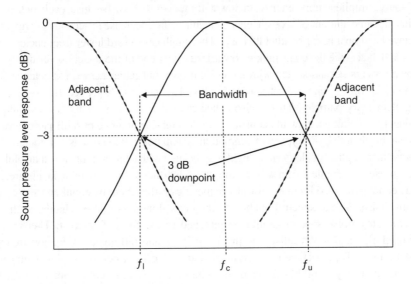

Figure 2.6 Frequency response of generic constant-percentage bands; the horizontal frequency scale is logarithmic

single smallest bandwidth. For example, for a pure tone of 94 dB at 1,000 Hz (typically used for calibration of sound level meters), the SPL would be 94 dB independent of the bandwidth. However, if the measured signal is 94 dB spread out over several octaves, the SPL per octave would be different from the SPLs in terms of 1/3-octave band levels, depending on the frequencies of interest. In general, SPLs in different frequency bandwidths can be calculated using the following equation:

$$SPL_1 = SPL_{band} - 10\log(BW)$$ (2.9)

where SPL_1 is the SPL in a 1 Hz bandwidth, SPL_{band} is the SPL in the bandwidth of interest, and BW is the bandwidth of interest $(f_u - f_l)$.

2.4.1   Noise by Color

Reference noise sources are sometimes labeled by color to use an optical analogy for the weighting of the frequency data. The most common colors associated with reference sound sources are white, pink, brown, red, and gray. White noise is characterized by equal energy with frequency on a linear scale. A white noise spectrum shape would be a horizontal line if the frequency scale is linear. This is analogous to the optical analogy as white light has equal light energy for all visual frequencies.

Pink noise is characterized by equal energy per octave band. This would result in a horizontal line for its frequency response if the frequency scale were logarithmic, which is typically the type of scale that is used for illustrating sound spectra. As, with octave bands, the frequency bandwidth of each successive band is twice that of the subsequent one, a pink noise source implies more energy in lower-frequency bands, because each octave band contains the same amount of acoustic energy, resulting in a decrease in acoustic energy of 3 dB per octave band on a linear bandwidth basis. This emphasis toward lower frequencies lends to the "pink" label, since the color pink is toward the lower end of the optical spectrum.

Brown noise is not associated with the color brown, but rather named for botanist Robert Brown (1773–1858), who is best known for his observation of so-called Brownian motion among elementary particles. This is also called red noise as it has a stronger low-frequency component than pink noise, with acoustic energy decreasing at a rate of 6 dB per octave band on a linear bandwidth basis, typical of ventilation system noise. Red noise is differentiated from pink noise in the optical analogy as red light is associated with a lower frequency than pink light.

Gray noise is weighted similarly to the equal-loudness curves shown in Figure 2.3 to provide equal perceived loudness for all frequencies rather than the equal energy for white and pink noise. As the spectral shapes of the equal-loudness curves change with sound level, true gray noise spectra should be matched to the SPL of interest. There are also lesser-used blue and violet noise, the inverse of pink and red noise, with acoustic energy increasing by 3 dB per octave for blue noise and by 6 dB per octave for violet noise, thus having more energy at higher frequencies. The blue and violet colors are used for these noise sources as they are associated with the upper end of the visual spectrum.

Of all these, white and pink noise are used most often as reference sources.

2.5   Common Sound Descriptors

2.5.1   Environmental Descriptors

The magnitude of sound energy is described by many mathematical derivations, so it is important that these descriptors are consistent and relevant to the situation being described. SPLs are rarely constant with time so these descriptors have been developed to rate sound levels for acceptability in different environments. In each case, the time duration associated with each descriptor must be identified and must be consistent with those of other descriptors to be relevant.

The most common sound energy descriptor is the equivalent level, designated as L_{eq}. L_{eq} is mathematically defined as:

$$L_{eq(t)} = 10 \log \left[\frac{1}{t} \int_0^t \frac{p^2}{p_{ref}^2} \, dt \right] \tag{2.10}$$

where $L_{eq(t)}$ is the equivalent level over the time period t, p is the acoustic pressure, and p_{ref} is the reference acoustic pressure (the threshold of hearing, or 20 µPa).

Equation (2.10) is an energy-average calculation of SPL over a prescribed period of time. This is a logarithmic average, so it tends to weight higher levels more than lower ones, making the values considerably different from linear averages when there is a wide variation in SPLs during the time period of interest. The linear average will only be the same as the L_{eq} when SPLs during the time period of interest do not vary. It is critical that the time period of interest is clearly stated for these values.

A descriptor based on L_{eq} that is widely used for rating public noise exposures is L_{dn} (sometimes written as DNL). L_{dn} is the day-night equivalent level (usually in dBA), which is based on a 24-hour L_{eq} calculation using 1-hour L_{eq} values and adding 10 dB to all hourly values occurring between 10:00 pm and 7:00 am to account for the added sensitivity to noise during normal sleeping hours. The L_{dn} will always be higher than the $L_{eq(24)}$ because of the 10 dBA night-time penalty. If the sound level in an area is constant during the entire 24-hour period of interest, the L_{dn} will be 6 dBA higher than the $L_{eq(24)}$.

A variation on L_{dn} is the L_{den} (sometimes called CNEL for community noise equivalent level). L_{den} is the day-evening-night equivalent level, and is calculated in the same way as the L_{dn} but with 5 dB added to all hourly L_{eq} values occurring during the evening hours of 7:00 pm to 10:00 pm and 10 dB added to hourly L_{eq} values between 10:00 pm and 7:00 am.

L_{eq} and L_{dn} are the most common descriptors used to rate sound exposures, but there are many other descriptors used for various purposes. One of those is the percentile level, L_n, where n is any number between 0 and 100, non-inclusive. L_n is the level exceeded $n\%$ of the time period of interest. As an example, L_{10} is the level exceeded 10% of the time period of interest. L_{50} is the level exceeded 50% of the time (the median level) and L_{90} is the level exceeded 90% of the time. L_{90} is often used to describe the residual background sound level in an area because it eliminates any transient events from its calculation. The L_1 value is an indication of the maximum level occurring during a period of interest and the L_{99} is an indication of the minimum level. L_n values are statistical and not averaged, so they are only relevant to the specific environment described.

A comparison between percentile levels gives an indication of the variation of sound levels during the period of interest. A difference between L_{10} and L_{90} of more than 15 dB indicates a significant amount of signal variation. L_{eq} is usually between L_{10} and L_{50} values for moderate variations (between 5 and 15 dB). When there is little variation in sound levels (less than 5 dB difference between L_{10} and L_{90}), L_{eq} will be close to the L_{50} value. When there is a significant amount of signal variation, L_{eq} will be close to or exceed L_{10} for the period.

L_{eq} values for the same reference time periods can be combined for the same or different sources but, due to their statistical bases, L_n values cannot be combined with other values. L_{eq} values (at the same location) for numerous sources can be combined using the following equation:

$$L_{eq(total)} = 10 \log \left[\sum_1^n \left(10^{\frac{Leqn}{10}} \right) \right] \qquad (2.11)$$

where n is an integer corresponding with an individual source and L_{eqn} is the equivalent level of each individual source at the same receiving/listening location. In this equation, each individual source contribution is reduced to its raw pressure value and the pressures are combined before using the logarithmic operation on them. This logarithmic addition is the appropriate method for combining dB values from different sources at the same listening location, rather than performing a linear arithmetic addition. Table 2.2 offers a simple method to use for combining SPLs from unique sources at the same listening location. This method is a simplification of using equation (2.11) and, if levels are combined two at a time, the resultant total value will be within 1 dB of the result calculated using equation (2.11).

Therefore, combining two sources having identical SPLs (individually) at the same location would result in adding 3 dB to the level attributed to one of the sources. This comes from the properties that the log to the base 10 of 2 is 0.3 and the logarithm of two multiplied values is the same as the log of the sum of the logs of each value individually (equation 2.1b), so doubling the number of sources would add 10 log(2) or 3 dB. In mathematical terms,

Table 2.2 Combining sound pressure levels at the same location (a simplification of equation 2.11)

dB difference between levels being combined	dB added to higher value
0 or 1	3
2 or 3	2
4 to 9	1
10 or more	0

$$10\log(2a) = 10\log(2) + 10\log(a) = 10\log(a) + 3$$

Another significant point from Table 2.2 is that, if the SPLs from two sources at the same location differ by 10 or more dB, combining the two would result in the same SPL as that associated with the louder source. Another way to look at this situation is that introducing a new source with a SPL 10 or more dB less than that of another existing source at a specific location would not raise the SPL at that location.

Here are some examples of combining SPLs using Table 2.2:

$$60\,dB + 60\,dB = 63\,dB$$

$$60\,dB + 62\,dB = 64\,dB$$

$$60\,dB + 65\,dB = 66\,dB$$

$$60\,dB + 70\,dB = 70\,dB$$

$$60\,dB + 80\,dB = 80\,dB$$

Since 0 dB does not correspond to a pressure value of 0 (but rather the pressure associated with the threshold of hearing), 0 dB + 0 dB would follow the same guidelines of Table 2.2 and result in an SPL of 3 dB.

A variation on this is for the situation of having n identical sources being combined at the same location, for which the total SPL would be:

$$L_{eq(total)} = L_{eqi} + 10\log(n) \tag{2.12}$$

where L_{eqi} is the equivalent SPL associated with one of the identical sources being combined. Therefore, a combination of 10 identical sources at the same location would result in the addition of 10 dB to the SPL at the listener. As a 10 dB SPL increase has been accepted as doubling the perceived loudness, 10 identical sources would sound twice as loud as a single one of those sources operating at the same location. Along the same lines and as mentioned earlier, doubling the number of identical sources at the same location results in an addition of 3 dB to the level at the listener. As a 3 dB increase is generally regarded as a just notice-able increase in sound level, doubling the number of identical sources at the same location would result in a just noticeable increase and a 10-fold increase in the number of identical sources at the same location would result in a doubling of loudness. These are key points to keep in mind when reviewing the sound level data in this book and other references.

Table 2.3 shows examples of dB values associated with perceived loudness changes compared with power changes. For this reason, the use of percentages when referring to changes in SPLs is discouraged as ratings can easily be confused.

Averaging L_{eq} values for unique sources at the same location can be accomplished using an equation similar to (2.11), as follows:

$$L_{eq(average)} = 10\log\left[\sum_1^n (10^{\frac{Leqn}{10}})/n\right] \tag{2.13}$$

Table 2.3 Correlation between perceived loudness changes and power changes for sound levels

SPL change (dBA)	Sound power change (dBA)	Perceived loudness change
3	Double	Just noticeable
5	3×	Clearly noticeable
10	10×	Twice
20	100×	4×
−10	0.1	0.5
−20	0.001	0.25

where n is an integer corresponding with an individual source and L_{eqn} is the equivalent level of each individual source at the same receiving/listening location.

When single acoustic events, such as aircraft or trains passing by, are of interest, the single event level (SEL, or sound exposure level) is sometimes used. SEL is a measure, in dB, of the total acoustic energy associated with a single acoustic event, compressed into a 1-second time interval. As long as the event's duration is longer than 1 second, the SEL will always be higher than any other descriptors, including the maximum instantaneous level.

2.5.2   Sound Propagation in Terms of Sound Levels

Divergence

Returning to the unimpeded sound propagation discussion in Chapter 1 (Section 1.3.1), point source sound propagation involves sound energy being dispersed as it is spread over an increasingly larger surface area of a sphere, $4\pi d^2$ (where d is the distance from the point source or center of the sphere, as illustrated in Figure 1.4). In terms of sound levels, the drop-off rate of SPLs from a stationary point source is related to the ratio of the distances at two locations with respect to the source. Since intensity (I, which is proportional to the square of the acoustic pressure) is power (W) per unit area and sound intensity level (L_I) is roughly the same as SPL (with pressure being force per unit area), the difference between SPLs at two locations would translate to the following (from equations 2.1a–c):

$$
\begin{aligned}
SPL_1 - SPL_2 &\approx L_{I1} - L_{I2} \\
&= 10\log\left(W/4\pi d_1^2\right) - 10\log\left(W/4\pi d_2^2\right) \\
&= \left[10\log(W) - 10\log\left(4\pi d_1^2\right)\right] - \left[10\log(W) - 10\log\left(4\pi d_2^2\right)\right] \\
&= 10\log\left(4\pi d_2^2\right) - 10\log\left(4\pi d_1^2\right) \\
&= \left[10\log(4\pi) + 10\log\left(d_2^2\right)\right] - \left[10\log(4\pi) - 10\log\left(d_1^2\right)\right] \\
&= 10\log\left(d_2^2\right) - 10\log\left(d_1^2\right) \\
&= 10\log\left(d_2^2/d_1^2\right) = 20\log\left(d_2/d_1\right)
\end{aligned}
\tag{2.14}
$$

Equation (2.14), also known as the inverse square law, implies that the drop-off rate of SPL from unimpeded divergence from a stationary point source is 6 dB per doubling of distance from the source. If the source is a line source, as illustrated in Figure 1.5, the divergence propagation pattern is cylindrical with a surface area of $2\pi d$ (where d is the distance to the source). Using this surface area in a similar derivation to that shown for equation (2.14), the SPL drop-off rate from a line source would be $10\log(d_2/d_1)$, or 3 dB per doubling of distance from the source.

Other factors in addition to divergence affect sound propagation outdoors, as mentioned in Chapter 1, and these include atmospheric absorption, atmospheric refraction, diffraction around barriers, and ground effects. The influence of atmospheric absorption is shown in Figures 1.12 and 1.13, and the influence of ground effects is shown in Figure 1.18.

Refraction and Diffraction

As mentioned in Section 1.3.4, atmospheric refraction effects can significantly affect outdoor sound propagation. The extreme conditions of temperature inversion and temperature lapse can account for more than a 20 dB difference in SPLs from a loud source at a distance of 300 m or more. Variations in wind profiles can have similar effects.

The diffraction effects illustrated in Figure 1.8, when translated to dB reductions, are 5 dB when the line of sight is broken between the source and listener, 5–7 dB just below the line-of-sight, and 8–12 dB well into the shadow zone, all for the mid-frequency range of 500–2,000 Hz and within 100 m of the barrier. Higher reductions are possible at higher frequencies and lower reductions can be expected at lower frequencies. In no case will a barrier provide more than 15 dBA of noise reduction, independent of its composition, due to diffraction effects. A complete enclosure would be required for the source if more than 15 dBA of sound reduction is needed.

Reverberation

As introduced in Section 1.3.3, reverberation is the persistence of sound energy within an enclosed space caused by repeated reflections off all the walls in the space. A common measure of reverberation is reverberation time (denoted RT_{60}), which is the time (in seconds) it takes for the SPL in a room to diminish by 60 dB after a sound source is deactivated. RT_{60} can be calculated using the Sabine equation (named for American physicist Wallace Clement Sabine [1868–1919]), as follows.

$$RT_{60} = 0.161V / A \,(\text{in s}) \tag{2.15}$$

where V is the volume of a space in m^3 and A is the total acoustical absorption of all room surfaces in units of metric sabins:

$$A = \Sigma \alpha_i S_i \tag{2.16}$$

where α_i is the absorption coefficient (a unitless value between 0 and 1 related to the percent absorption effectiveness) and S_i is the surface area of the associated wall material.

The value A varies with frequency and so RT_{60} varies with frequency. As equation (2.15) implies, RT_{60} can be reduced by reducing the size of a room or by adding absorptive finishes. RT_{60} values can dictate the ideal acoustic use for a space, with low mid-frequency (500–1,000 Hz) RT_{60} values (less than 1 s) being best for rooms requiring good speech intelligibility (such as lecture rooms) and higher mid-frequency RT_{60} values (greater than 2 s) being best for rooms requiring the blending of music (such as classical or organ music venues).

Sound Fields

The monitoring of SPLs can be affected by the sound field in which the measurement is being performed. This can also affect the validity of the measurement. The basic sound fields are characterized as near, far, free, direct, reverberant, and diffuse.

The near field is within roughly a quarter-wavelength of the lowest frequency of interest away from a source. SPLs can vary widely and unpredictably in the near field and therefore measurements should not be taken very close to a source or large reflective surface. Far, free, and direct fields are all regions in which SPLs drop off in accordance with the inverse square law for point sources. The far field is so named in order to differentiate it from the near field, in which there is no consistent drop-off rate. The free field is named to imply that there are no obstructions in the path of the sound propagation. Free fields are usually referenced for outdoor sound propagation. The direct field is named to imply that the measured sound wave is only that which is traveling directly from the source to the listener, with no reflections off local surfaces to change the level or properties of the sound wave. Direct fields are typically used for relatively close indoor measurements to establish reference levels for specific sources.

Reverberant and diffuse fields are the same, each defining an area in an enclosure in which reverberation is prevalent enough to create regions in which there is no variation in SPL with distance from a source. The location in a room at which the transition takes place between the direct and reverberant fields is known as the critical distance, d_c, mainly dependent upon the absorption within the room and equal to $0.141\sqrt{A}$, where A is defined in equation (2.16).

A distinction should be made between the terms ambient and background sound levels since they are often used interchangeably. The ambient sound level is associated with the entire acoustic environment including all sources contributing to that environment while the background sound level is associated with all sound sources except for a defined source of interest. Comparing the ambient sound level to the background sound level therefore provides a comparison between levels with and without a source, permitting one to determine the potential effect of adding a source of interest into an acoustic environment. The terms ambient and background would only be interchangeable if there is no need for such a comparison.

References

[1] Yost, W.A. *Fundamentals of Hearing*, [5]th edn. Bingley, UK: Emerald Group Publishing Limited, 2008.

[2] Wiener, F.M. (1947). "On the diffraction of a progressive sound wave by the human head." *Journal of the Acoustical Society of America*, 19(1): 143–146.

[3] Suzuki, Y. and Takeshima, H. (2004). "Equal-loudness-level contours for pure tones." *Journal of the Acoustical Society of America*, 116(2): 918–933.

[4] International Organization for Standardization. *ISO 226:2003. Acoustics – Normal Equal-Loudness-Level Contours.* Geneva: ISO, 2003.

[5] Fastl, H. and Zwicker, E. *Psychoacoustics: Facts and Models*, 3rd edn. Heidelberg, Germany: Springer, 2007.

[6] Hellman, R.P. and Zwislocki, J. (1961). "Some factors affecting the estimation of loudness." *Journal of the Acoustical Society of America*, 33(5): 687–694.

[7] Acoustical Society of America. *ANSI S1.42-2001 (R2011). Design Response of Weighting Networks for Acoustical Measurements.* New York, NY: American Institute of Physics, 2011.

[8] International Organization for Standardization. *ISO 7196: 1995. Acoustics – Frequency-weighting Characteristic for Infrasound Measurements.* Geneva, Switzerland: ISO, 1995.

[9] Stevens, S.S. (1972). "Perceived level of noise by Mark VII and decibels (E)." *Journal of the Acoustical Society of America*, 51(2, Part 2): 575–601.

3

Sound Perception

3.1 Introduction

The final step in laying the foundation for an understanding of the effects of sound on people is an explanation of the human hearing mechanism. The anatomy of the human hearing mechanism is introduced, along with an explanation of the different ways in which sound waves become converted into electrical signals that are interpreted by the brain as sound. Normal hearing methods and hypersensitivities are addressed to clarify why some sounds bother some people more than others. This chapter provides an overview of the aspects of sound perception that have been explored by researchers in recent history.

3.2 Human Hearing Apparatus and Mechanism

The anatomy associated with the human hearing mechanism provides a remarkable transduction system for the efficient conveyance of sound energy to the brain. Our sense of hearing has been crucial to our survival, mainly for communication and to react to impending danger. Modern society has minimized the need for a survival reaction in humans, but hearing is still vital to the survival of most other species.

A transducer is a device that converts energy from one form into another. The human hearing mechanism performs multiple transductions, converting acoustic energy into mechanical energy and then into electrical energy before it is interpreted as sound by the brain. The anatomy that accomplishes these transductions is typically divided into three regions – the outer, middle, and inner ears – each being separated by the type of energy conveyed. The outer ear channels acoustic energy to the middle ear, which channels transduced mechanical energy to the inner ear, which in turn transduces mechanical into electrical energy before it is sent through the auditory nerve to the brain for processing.

The Effects of Sound on People, First Edition. James P. Cowan.
© 2016 John Wiley & Sons, Ltd. Published 2016 by John Wiley & Sons, Ltd.

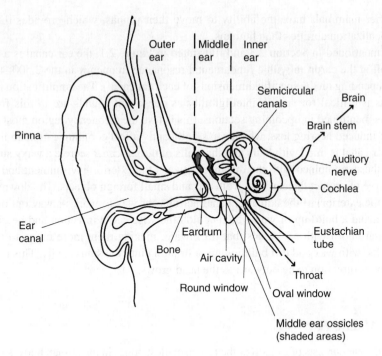

Figure 3.1 Schematic diagram of the human hearing mechanism

Figure 3.1 shows a schematic diagram of the entire system. Mammals in general have similar systems with differing component shapes and sizes, which dictate their differing frequency sensitivities.

3.2.1 Outer Ear

The outer ear consists of the externally exposed section of the hearing mechanism, from the pinna (also called the auricle) to the eardrum (also called the tympanic membrane). The primary purposes of the outer ear are for localization, resonance, and protection. Sound energy is first introduced to the hearing mechanism through the pinna, the visible external part of the hearing mechanism resembling a horn shape, with folds to channel sound into the ear canal and assist in source localization. The tragus, a small flap of cartilage on the outside of the ear canal (not shown on Figure 3.1) also assists in this localization process. The head and shoulders affect sound fields as well through shielding and reflection, especially below 1,500 Hz [1].

Localization of sound sources in the horizontal plane is effectively achieved by processing the difference in arrival times of signals in each ear. As our ears are each mounted at the same elevation, however, vertical localization cannot be accomplished by the same means. The shape of the pinna helps in vertical localization but still cannot provide the level of localization afforded by horizontal time difference processing. Humans therefore have better localization abilities in the horizontal than in the vertical directions.

Most other mammals have the ability to move their pinnas, which provides them with better localization abilities than humans.

As is mentioned in Section 2.3 and illustrated in Figure 2.1, the ear canal is a cylinder closed off at the eardrum with a fundamental resonance frequency in the 2,500–4,000 Hz range (depending on the specific dimensions of each ear canal). The amplification resulting from this is critical for speech intelligibility as consonant sounds are in this frequency range. As mentioned in upcoming sections, this is also the frequency region most affected by noise-induced hearing loss, which can significantly affect communication ability.

The ear canal is lined with hairs and the walls of the ear canal secrete a waxy substance. Both of these, combined with the distance between the exterior environment and the eardrum, serve to protect the eardrum from dirt and small foreign objects. The downward tilt (toward the exterior) of the adult ear canal permits ear wax to work its way out of the ear canal to avoid a build-up of wax which could otherwise cause infection and hearing loss. Children are born with ear canals that tilt inward, making them more prone to infection than adults, with wax and dirt building up or draining into the sinus cavity. This tilt gradually changes direction to be outward as the head grows.

3.2.2   Middle Ear

The middle ear consists of a chain of the three smallest bones in the human body, suspended in an air cavity with the support of ligaments and two muscles, attached to the eardrum on one end and the oval window of the cochlea on the other end. The three bones in this chain – the malleus, incus, and stapes (also known as the middle ear ossicles) – carry vibration from the eardrum to the oval window, and perform as an amplifying device to account for the diminishment of the acoustic signal between the outer and inner ears.

Also considered part of the middle ear is the Eustachian tube (named for the Italian anatomist Bartolomeo Eustachi [c. 1500–1574]), which connects the middle ear cavity with the upper throat (nasopharynx) for pressure equalization. Under normal conditions, the Eustachian tube is sealed at the throat, except when swallowing. When atmospheric pressure changes, as occurs when we change altitude, the pressure cannot change in the middle ear cavity until the Eustachian tube is open to the outdoors. A sensation of fullness at the eardrum is common until the pressure is equalized with the external environment by swallowing and opening the Eustachian tube.

One obstacle that needs to be overcome in this part of the system is the impedance mismatch between the air in the middle ear and the fluid in the inner ear. As mentioned in Chapter 1, the acoustic impedance of a medium affects the flow of acoustic energy through the medium and into another medium. If there is a significant mismatch in acoustic impedance between media, the flow of energy will be impeded and possibly reflected back at the medium interface. The characteristic acoustic impedance is a function of the density of a medium and it is inversely proportional to the surface area in the direction of energy flow. For this reason, a sharp discontinuity in cross-sectional area of energy flow in the same medium can cause the reflection of sound back into the medium. An example of this is the closed–open system illustrated in Figure 1.11. This can be avoided by a gradual change in cross-sectional area in the sound path, as is the case for loudspeaker horns and the pinna of the outer ear. This is why hearing abilities are significantly improved by having pinnas on our heads. Without pinnas, there would be a sharp discontinuity between the environment outside our ear canals and the ear canals themselves, because the opening to the ear canal would be a hole flush with the side of our heads, as is the case for fish and reptiles. This would cause an impedance mismatch, which limits hearing response.

After sound pressure waves travel down the ear canal, they set the eardrum into sympathetic vibrations, converting the sound energy into mechanical energy. These vibrations are carried by the ossicular chain of the malleus, incus, and stapes. These bones are named for the Latin versions of hammer, anvil, and stirrup, respectively, because of their shapes. The malleus is attached to the back of the eardrum and the stapes is attached to the outside of the oval window (on the cochlea). As the medium behind the oval window in the inner ear is a fluid, the impedance mismatch between the air and the fluid will diminish the signal if it is unaltered between the eardrum and the oval window. The eardrum is roughly 10–18 times larger than the oval window, which causes an amplification of roughly 25 dB between the eardrum and the oval window as the same force (at the eardrum) is being applied over a reduced area at the oval window. However, the impedance mismatch between the eardrum and the oval window requires an amplification of roughly 60 dB for the efficient conveyance of the acoustic signal to the inner ear. This difference is accounted for by the lever action of the ossicular chain; however, these amplifications are frequency-dependent, with the greatest amplifications occurring in the same range as ear canal resonance, and can explain the shape of the equal-loudness curves shown in Figure 2.3.

The two muscles in the ossicular chain are the tensor tympani and stapedius muscles. These muscles provide protection for the hearing mechanism by limiting the amount of energy being conveyed to the inner ear when sound exposure levels are excessive. Each of these muscles functions in different ways to reduce sound pressure levels at the oval window by up to 30 dB, mostly for frequencies below 2,000 Hz. The tensor tympani muscle is connected to the malleus and the eardrum. Contraction of the tensor tympani muscle stiffens the eardrum and this is a neural reflex activated each time one is startled, or when one eats and speaks to reduce the high internal sound levels generated by chewing or vocalizing. As the nervous system detects when a person will eat and speak, it is prepared to activate this acoustic reflex at the time activity begins. It is not as effective for unpredictable external sources, though, for which the stapedius muscle provides protection after a minimum delay of 100 milliseconds (ms [1 ms = 10^{-3} s]).

The stapedius reflex causes the stapedius muscle (the smallest muscle in the body) to stiffen the ligaments supporting the stapes, thereby reducing the transmission of sound energy. As this acoustic stapedius reflex is delayed, a high-level (greater than 80 dB above each person's hearing threshold), unexpected, impulsive signal (such as an explosion or gunfire) would not benefit from this protection mechanism, thereby exposing the hearing mechanism to the full exposure level. Although not recommended here, some researchers have suggested that humming before shooting a gun may activate the acoustic reflex from vocalization and therefore avoid the 100 ms delay [2].

Along the same lines, experiments have been performed that show a clear reduction in noise-induced hearing loss from exposure to gunfire after a pre-exposure to a 1,000 Hz tone has activated the acoustic reflex [3]. In any case, proper hearing protection should always be worn when using firearms to protect against noise-induced hearing loss. Considering that typical firearms can generate sound pressure levels above 160 dBA at the shooter's ear, the 20–30 dBA attenuation from the acoustic reflex will still leave the ear vulnerable to sound pressure levels above damaging limits. This will be discussed further in the section on noise-induced hearing loss in Chapter 4.

Otosclerosis is a hereditary condition in which there is a rigid connection between the stapes and the oval window, thus limiting the transmission of sound energy to the inner ear. This typically results in a low-frequency hearing loss, in contrast to the mid- and high-frequency losses associated with the sensorineural hearing loss conditions discussed later in this chapter and in Chapter 4. One method used for testing of otosclerosis is to startle a person with a burst of air at the opening to the ear canal and note whether the stapedius reflex occurs [4].

3.2.3   Inner Ear

The stapes vibrates the oval window, which is the entry point to the inner ear. The inner ear consists of the cochlea, a bone-encased spiral divided into three sections, the center section of which houses the key sensory organs that transduce the mechanical vibrations from the middle ear into electrical signals that are sent to the brain for processing.

Another part of the inner ear is not directly related to hearing. The semicircular canals above the cochlea regulate the vestibular or balance system. These are three fluid (endolymph)-filled looped ducts above the cochlea (see Figure 3.1) that sense head movements in different directions. Each loop is oriented 90 degrees with respect to any others to sense head motion in all directions. At the base of the semicircular canals near the oval window is the vestibule, housing the utricle and saccule, from which each has nerve bundles that combine with the auditory vestibular nerve to send signals regarding balance to the brain. The utricle senses horizontal movements and the saccule senses vertical movements of the head. Vertigo issues are usually associated with irregularities of the vestibule or semicircular canals of the inner ear. A common disease of the inner ear, labyrinthitis, can cause debilitating vertigo for short periods.

Figures 3.2(a–d) show an ever expanding view of the inner ear hearing apparatus. Figure 3.2(a) depicts a cross-sectional view of the cochlea, showing the interior of the spiral sections separated by membranes. As shown in Figure 3.2(b), the spiral duct inside the cochlea is divided into three sections – the scala vestibuli, scala media, and scala tympani. All of these sections are filled with fluids, the scala vestibuli and scala tympani with perilymph and the scala media with endolymph. The scala vestibuli extends from the oval window to the apex of the cochlea, known as the helicotrema. The scala tympani extends from the round window (a membrane on the outer wall of the cochlea just below the oval window, not connected to the stapes) to the helicotrema, where the scala tympani and scala vestibuli meet. The round window provides pressure equalization from the disturbed fluids and membranes between the oval window and round window. Because the fluids in the inner ear are incompressible, a disturbance initiated by the movement of the oval window at the entrance of the cochlea must be countered by movement of the round window.

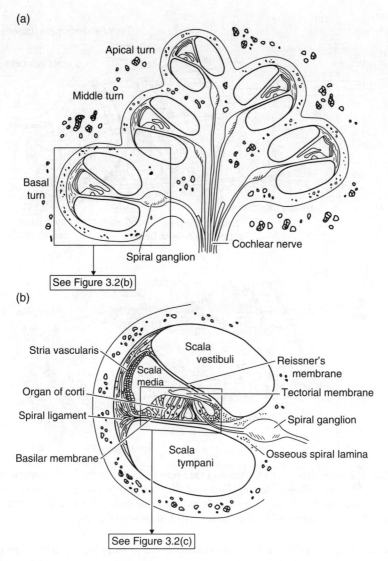

Figure 3.2 Schematic diagrams of: (a) a cross-section of the human cochlea; (b) a cross-section of one turn of the human cochlea

The scala media is the central section of the cochlear spiral, which houses the signal processing center of the cochlea in the organ of Corti, named for Italian physician Alfonso Corti (1822–1876), where the mechanical energy sent from the middle ear is transduced into electrical energy. In the outer wall of the scala media is the stria vascularis, which produces endolymph for the scala media and provides an oxygen-rich blood supply for the inner ear.

As Figure 3.2(a) shows, the organ of Corti has nerve endings that combine at the exit of the cochlea into the auditory nerve, which carries the electrical signals to the brain where they are interpreted as sound. The auditory nerve combines with the vestibular

(c)

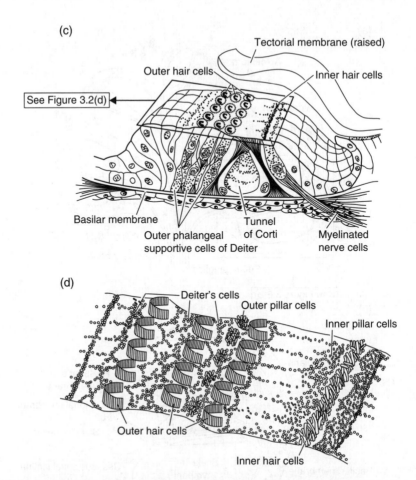

Figure 3.2 (*Continued*) (c) a cross-section of the organ of Corti; (d) the arrangement of hair cells in the organ of Corti. (From: *Hearing Loss and Tinnitus: Discussion Paper prepared for The Workplace Safety and Insurance Appeals Tribunal*, by John Rutka MD FRCSC (February 2013; revised July 2013) © Queen's Printer for Ontario, 2013. Reproduced with permission.)

nerve before it enters the brain, with the vestibular nerve carrying electrical signals associated with the balance information generated from the semicircular canals. This is the eighth of the 12 cranial nerves and is therefore called the auditory vestibular, or vestibulocochlear, nerve.

The organ of Corti is bounded by the basilar membrane on one side and the tectorial membrane on the other. The scala media section is bordered by the basilar membrane and the Reissner's membrane (named for German anatomist Ernst Reissner [1824–1878]), as shown in Figure 3.2(b). Figure 3.2(c) shows the organ of Corti in greater detail. The key transduction components of the organ of Corti are the hair cells, arranged as a single row of inner hair cells and three or more rows of outer hair cells. There are roughly 3,500 inner hair cells and 12,000 outer hair cells in the human cochlea. The inner hair cells are connected to nerve fibers that combine in the auditory nerve.

Both the inner and outer hair cells have small individual cells called stereocilia that are arranged in U and V shapes, as shown in Figure 3.2(d). The outer hair cells are supported by Deiters cells (named for German neuroanatomist Otto Deiters [1834–1863]). They are also surrounded by perilymph and in contact with the tectorial membrane. The inner hair cells are surrounded by supporting cells and are not in direct contact with the tectorial membrane. When the stapes vibrates the oval window at the entrance to the cochlea, longitudinal (back and forth, as in piston motion) waves are set up in the endolymph of the scala media along with the basilar and tectorial membranes on either side of the organ of Corti's hair cells. These waves result from the pressure differences in the liquids between the scala vestibuli and the scala tympani. Shearing movements between the basilar and tectorial membranes in response to these waves cause the hair cells to move, which in turn generate electrical signals that are transmitted through the auditory nerve to the brain. These electrical signals are generated by the ionic difference between potassium in the endolymph of the scala media and sodium in the perilymph of the scala vestibuli and scala tympani. As the outer hair cells are in direct contact with the tectorial membrane, they are displaced directly by the shearing movements. The inner hair cells are displaced by the movement of fluids through the chamber as they are not in direct contact with the tectorial membrane.

von Békésy's experiments [5,6] showed that travelling waves generated in the basilar membrane cause higher frequencies to be sensed toward the base of the cochlea and lower frequencies toward the apex near the helicotrema. The helicotrema is the only location along the cochlear spiral where the scala vestibuli and scala tympani combine and communicate with each other. Because of this, the pressure difference between these chambers is reduced at the helicotrema. A combination of this with the condition of the inner hair cells not being in direct contact with the tectorial membrane contributes to our reduced sensitivity to lower-frequency sounds [7].

There are two accepted methods to explain how the cochlea processes frequency data – the place and the temporal theories. The place theory assumes frequencies are discerned primarily by the location of the stimulated hair cells along the spiral cochlea, while temporal theory assumes that the rate of hair cell excitation stimulates the sensation of frequency. Both of these theories have been experimentally verified below 5,000 Hz, but the temporal theory can only be valid below 5,000 Hz, as that is the limit of processing speed for our auditory neural system [8].

The inner hair cells carry most of the sound information to the brain through afferent connections (from the hair cells to the brain) and the outer hair cells amplify and refine that information as it is fed back from the brain through efferent connections (from the brain to the hair cells). Roughly 5% of outer hair cells have afferent connections with the brain. Outer hair cells have the ability to expand and contract (known as motility), while inner hair cells do not have that capability. This is thought to be the basis for the phenomenon of otoacoustic emissions, whereby echoes of sounds introduced into the ear canal can be detected returning back from the cochlea roughly 10 ms later. This was first reported by Kemp in 1978 [9]. Otoacoustic emission measurements are now widely used as indicators of abnormalities in cochlear functions, because a normally operating cochlea will generate these emissions and the absence of those emissions can indicate an issue.

In addition to fine-tuning the auditory system, the outer hair cells add a layer of protection against high-amplitude sounds in addition to the middle ear muscle reflexes described earlier. The medial olivocochlear reflex uses the motility of the outer hair cells to stiffen the basilar membrane and thus reduce the signal levels sent to the brain [10].

3.2.4 Signal Processing in the Brain

The processing of the electronic signals carried from the auditory nerve to the brain is complex and beyond the scope of this book. Textbooks devoted to the hearing process can be consulted for details about this, but the basic effects of localization, precedence, and masking are described herein. In addition to the accepted path of signals from the auditory nerve through the superior olivary complex to the auditory cortex of the brain, however, there are neural pathways from the hearing mechanism to the amygdala, where fear and emotional responses are processed [11]. This is the basis for the psychological sound effects on people discussed in Chapter 5.

Localization

Localization in the horizontal plane, as mentioned earlier in the outer ear discussion, is mainly accomplished by processing the arrival time and sound level differences between sound waves at each ear, the binaural aspect of hearing which requires two functional ears to communicate signals to the brain. Arrival time differences are most important for lower frequencies and level differences are most important for higher frequencies, as sound waves are more affected by diffraction around the head at higher frequencies (due to their shorter wavelengths). As there is no arrival time or level difference when source locations vary in the vertical plane, this type of signal processing has little effect for localization of sources above, below, behind, and in front of a person. Localization in the vertical plane is mostly related to the pinna, and it has been shown that vertical plane localization is independent of monaural (hearing with one ear) or binaural (hearing with two ears) hearing [12,13]. The folds of the pinna provide a significant amount of information for localization of sounds, especially for frequencies above 6,000 Hz, as time delays as short as 20 microseconds (μs, or 10^{-6} s) caused by pinna reflections can be processed by the brain [14]. For middle and lower frequencies, the diffraction of sound around the head and shoulders assist with vertical localization [15,16].

The localization described here can result from single or multiple sound sources. Another aspect of localization from processing multiple signals is known as the precedence effect, originally discussed by Wallach et al. [17] and Haas [18]. Also known as the Haas effect, this results from the processing of multiple acoustic signals with short delay times between them. When the arrival time difference between two signals is between 1 and 50 ms, we hear a single resultant enhanced sound, but the sound source appears to be in the direction of the first arriving signal. This phenomenon occurs for delayed signals up to 10 dB higher than the first-arriving signal. This effect disappears when the delayed signal is at least 15 dB higher than the first-arriving signal. When the delay time between two signals is less than 1 ms, the source appears to be between the locations of the two sources [19]. This effect has been used in live concerts and recordings to manipulate the appearance of the locations of different sound sources. It also disappears when the delay between signals exceeds 50 ms, in which case we hear the two sounds distinctly as echoes, as mentioned in Chapter 1.

One other localization effect worth noting is the cocktail party effect, originally discussed by Cherry [20], which is the phenomenon by which we can focus attention on a specific sound source in the presence of other sources competing for our attention. Since Cherry's original experiments, many researchers have attempted to explain this phenomenon, most recently through targeted sound masking [21] and brain scan evaluations [22], but no conclusive studies have been performed to explain the phenomenon.

Masking and Audibility

Masking is the process of rendering one sound inaudible in the presence of another, which is tied to the way our auditory system processes sounds. Beginning with the experiments of Fletcher and Munson [23,24] and continued by Zwicker and his colleagues, [25] this processing has been shown to closely match the 1/3-octave band frequency scale down to the 250 Hz range, and the frequency bands associated with masking have been called critical bands for human hearing. These bands are similar in shape to those shown in Figure 2.6. Below 250 Hz, the bandwidth of critical bands is fairly constant at roughly 100 Hz, making the bands overlap for low-frequency processing. This has a role in our diminished sensitivity to low-frequency sounds, but it mainly plays a role in masking. The masking level is associated with the threshold of hearing under conditions of background noise, whereas the hearing threshold curves shown in Figures 2.2 and 2.3 are based on the condition of no background noise.

Critical bands are defined by the condition of tones being just audible in the presence of noise. If two tones are presented that have the same level and are within a single critical frequency band, we will hear a combination of those tones but not an increase in loudness. If those two tones are each in different critical bands, we will perceive the tones as clearly separate and louder. This is different from localization processing discussed earlier; this addresses the audibility of sounds rather than sensing their direction of origin. When a tone is just audible, its sound power is equal to the sound power associated with the background noise in the relevant critical frequency band.

When two tones have exactly the same frequency, they are combined into a single louder tone. When there is a slight difference in frequency between the two tones, a pulsing sound is heard at the frequency of the difference between those two frequencies. These are known as beats. For example, if a 500 Hz tone generates the same sound pressure level as a separately presented 502 Hz tone, the hearing mechanism will process a beat frequency of 2 Hz which will distort the sound into a pulsing 500 Hz tone at a rate of twice per second. As the frequency of the second tone is increased, the beating perception will transition into a harshness in the quality of the sound. As the frequency of the second tone is increased further, the harshness transitions into a more pleasant blending of the frequencies. It is at this blending point that the limit of the critical band has been reached.

As von Békésy reported [5,6], low-frequency sounds excite the entire length of hair cells along the cochlear spiral, while higher frequencies mainly excite hair cells near the entrance of the cochlea at the oval window. For this reason, low-frequency sounds tend to mask higher-frequency sounds, while high-frequency sounds are not effective in masking low-frequency sounds. Two tones that are significantly different in frequency exhibit minimal masking.

The general statement in Chapter 2 of a 3 dB increase constituting a just-noticeable increase has been adopted by the environmental noise community, but it should be noted that the audibility of sound depends on both frequency and the masking level of background sound.

Another phenomenon associated with background noise is the Lombard effect (named for French otolaryngologist Etienne Lombard [1869–1920]), which is an involuntary increase in voice level and change in voice characteristics (such as pitch, vocal rate, and enunciation) when background noise levels are elevated. The specific conditions under which the Lombard effect is triggered vary with the individual and are different for males and females [26]. This phenomenon is not limited to humans as it occurs in most species [27].

3.2.5   Vestibular System

The auditory vestibular nerve carries electrical signals to the brain from both the cochlea (for sound) and the vestibule and semicircular canals (for balance through the vestibular system). As for the main hearing organs of the cochlea, the central region of the semicircular canals is encased in a membrane, filled with endolymph, and lined with hair cells that generate electrical signals sent along the auditory nerve to the brain for processing. The region between the membrane and the outer bone casing is filled with perilymph to promote the same electrical polarization from hair cell stimulation as occurs in the organ of Corti. The utricle and saccule in the vestibule are also lined with hair cells. These hair cells are most sensitive to head movements and high levels of low-frequency sounds. Because of this, high noise levels can affect the vestibular system. However, as discussed in Chapter 4, extremely high sound levels are required to have a repeatable effect. Even in that case, only asymmetrical exposures have been shown to potentially cause lasting vestibular problems [28], as the vestibular system tends to adjust to these conditions by compensating in each of the two sets of semicircular canals. Although sopite syndrome, an illness characterized by symptoms typically associated with fatigue and motion sickness, has been linked partly with the vestibular system [29,30], sound exposure has not been suggested as a stimulus for this condition.

Head trauma injury or abnormal physiology can result in the Tullio phenomenon (named for Italian biologist Pietro Tullio [1881–1941]), for which sounds of varying intensities can cause the sensation of vertigo and involuntary eye movements, known as nystagmus [31]. This is thought to be caused by a third membrane window in addition to the oval and round windows in the inner ear, which displaces endolymph fluids in the semicircular canals. This fluid movement results in hair cell stimulation in the semicircular canals, which send signals to the brain that are interpreted as dizziness. This third window does not exist in normal anatomy, and the vestibular system normally operates independently of the auditory system even though the auditory vestibular nerve carries information to the brain from both systems.

Conditions that cause symptoms associated with the Tullio phenomenon are an opening in the bone above the semicircular canals, known as superior canal dehiscence [32], and a tear in the round or oval window, known as perilymph fistula. These openings can cause leaks of fluids from the inner ear and pressure sensitivities to sound energy conducting through the middle and inner ear regions. As perilymph fistula is a membrane tear, it often heals without medical intervention. This is not the case for superior canal dehiscence, because this is an opening in bone.

3.3   Alternate Sound Perception Mechanisms

The method of sound energy transduction described above is the most efficient method by which sound is perceived; however, there are several other documented ways in which sound energy can be sensed by the brain. These are bone conduction, cartilage conduction, tinnitus, and electromagnetic hearing.

3.3.1   Bone Conduction

Bone conduction (documented in the 19th century by Thomas Barr [33]) bypasses the outer and middle ears, causing vibrations of the skull to be transmitted directly to the cochlea. For a person speaking, bone conduction significantly changes the frequency response of subjective voice perception. As mentioned earlier, the middle ear muscles (especially the tensor tympani muscle) react to one's own speech. The combination of this acoustic reflex with bone conduction causes one's own voice to sound different than it does to others.

There is roughly a 60 dB difference in thresholds for bone conduction compared with hearing thresholds for air-conducted sound traveling through the outer and middle ears, in addition to skull resonances that can cause distortions in bone-conducted signals. Bone conduction tests are useful for determining whether a hearing loss is conductive (due to a problem in the middle ear) or sensorineural (due to a neural issue between the cochlea and the brain). There are also bone conduction-based hearing aids that can be effective for people with conductive hearing loss.

3.3.2   Cartilage Hearing

Recent research in Japan [34] has revealed the phenomenon of cartilage conduction, whereby a transducer is attached to the tragus area of the pinna near the ear canal opening and used as a hearing aid. This method is especially effective for frequencies below 3,000 Hz, without the common limitations of bone conduction hearing aids because cartilage hearing is mostly unrelated to bone conduction, generating sounds up to 35 dB higher than those transmitted through the ear canal alone [35].

3.3.3   Tinnitus

Tinnitus is a condition affecting up to 20% of the population [36] in which sounds (most typically ringing or buzzing) are heard without an associated external stimulus. Although the cause and mechanisms involved in tinnitus have been studied for many years, there is no consensus on either its cause or a cure. Tinnitus can range from a mild annoyance to being debilitating.

It is agreed that in most cases tinnitus is caused by damage to the neural network of the auditory system. An over-exposure to noise often results in tinnitus as the hair cells in the inner ear are overextended, but the tinnitus typically subsides after short periods as long as the person is removed from the area of high noise exposure. If the hair cells are permanently damaged, adjacent hair cells can overcompensate for the reduced activity from the damaged cells, causing an increase in neural activity and resulting in hearing the phantom sounds associated with tinnitus. Although these extra signals are often canceled by a feedback loop in the limbic region of the brain, a compromise in that system can result in a permanent chronic tinnitus condition [37,38].

Another potential cause of tinnitus is the aging process, as the neural network in the brain breaks down, although noise exposure is a primary cause for younger individuals. Tinnitus can accompany the condition known as hyperacusis, for which there is heightened sensitivity to all sounds (discussed further in Section 3.4.1). Tinnitus is also frequently reported by people diagnosed with dementia [39]. These cases seem to trace back to a dysfunction in the auditory processing center of the brain.

There is still no consensus on the cause of tinnitus. Some researchers have linked tinnitus with damage to hair cells but others place the source along the auditory vestibular nerve or in different centers of the brain. Research is ongoing in this area to identify the definitive cause. Although there are many treatment options available for tinnitus, there is no consensus that any of the methods are effective [40]. These include medications, electrical stimulation, masking sound introduction, biofeedback, hearing aids, counseling, acupuncture, and hypnosis.

Most tinnitus is perceived as high-frequency sound. A special case of tinnitus, known as hums, results (most often) in the sensing of low-frequency noise [41] when there is no obvious external acoustic source. Hum cases have surfaced over the past 40 years worldwide, some involving obvious acoustic sources and others remaining unsolved. Solved and unsolved hum cases are discussed in Chapter 6.

3.3.4   Electromagnetic Hearing

Although electromagnetic energy is a different form (and different frequency range) from acoustic energy, acoustic energy is transduced into electromagnetic energy in the inner ear, and electromagnetic, rather than acoustic, energy is the final stimulus for the sensation of hearing in the brain. It is well known that other animals can sense the naturally occurring electromagnetic fields around the Earth. As an example, birds are known to sense naturally-occurring geomagnetic fields for their use in migration [42]. The premise of humans being able to hear electromagnetic energy has generally been dismissed, but some scientists have proved there is the potential for that to occur.

Experiments prior to the 1960s proved that some electromagnetic signals can generate hearing sensations in the brain for electromagnetic sources close to the body and ear [43,44]. Another experiment in the 1970s involving electrodes connected directly to the body induced skull vibration which resulted in the subjects hearing sounds by bone conduction [45].

Although the hearing mechanism associated with these phenomena was not clearly understood, experiments in the 1960s showed that hearing through mechanisms outside the ear was associated with the sensation of hearing electromagnetic waves, beginning with the work of Allan Frey [46]. In addition to experiments in a controlled laboratory environment, Frey demonstrated the potential for people to hear electromagnetic signals from antennas up to several thousand feet away from subjects. In all cases, the sounds appeared to be originating inside the head or directly behind the head (independent of the orientation of the head to the source). This phenomenon of hearing electromagnetic signals has been labeled as microwave hearing. His later experiments agreed with other studies indicating that the electromagnetic fields induce skull vibrations that are carried by bone conduction to the cochlea [47].

The mechanism by which people can sense these signals is still not conclusive at this point, but Frey's experiments showed no difference with head position or in subjects having otosclerosis, which would rule out any associations with the outer or middle ears. The current research consensus states that the minute, rapid thermal expansion of soft head tissue (known as thermoelastic expansion) in the head resulting from electromagnetic field exposure generates signals that are sent by bone conduction to the cochlea for processing as sound [48–50].This phenomenon seems to be prevalent for people with good hearing sensitivities above 5,000 Hz, and does not appear to pose a health risk [51].

3.4   Hypersensitivities

Owing to the remarkability of the intricate workings in the human hearing mechanism, there is much room for imperfections. Slight abnormalities in any part of the intricate system can have dramatic effects on the results and experiences of the individual. There are many diseases associated with the hearing mechanism. These diseases can have significant quality-of-life effects, mainly associated with hearing ability or balance, or a combination of the two. An example is Ménière's disease (named for French physician Prosper Ménière [1799–1862]), for which there is excess fluid in the inner ear causing distension and rupture of the membranes separating the chambers of the cochlea (also known as endolymphatic hydrops). This can cause a mixture of the normally separated endolymph and perilymph fluids, which can lead to severe vertigo accompanied by hearing loss, tinnitus, and a feeling of pressure or fullness in the ear. Many of the symptoms of Ménière's disease are similar to those for the Tullio phenomenon discussed earlier, although the causes are different.

In addition to diseases of the auditory system, there is a segment of the population that is hypersensitive to sound and electromagnetic fields, which can have debilitating effects. These are briefly discussed in the following sections.

3.4.1   Hyperacusis/Misophonia

Hyperacusis is the state of being overly sensitive to sound. Normal sounds that would not bother most people can be intolerable to someone with hyperacusis. Although hyperacusis has also been associated with misophonia (the hatred of sound), misophonia is associated with the psychological aspect while hyperacusis has physiological associations as well. A variation of misophonia is phonophobia, or the fear of sound. Although both of these result in the need to avoid certain sounds, the key differences between misophonia and phonophobia are their associated reactions. Misophonia often starts in childhood and can severely restrict the lifestyles of those who experience it [52]. Those who experience misophonia generally have no tolerance for certain sounds, such as those associated with others' eating, coughing, or breathing, while those who experience phonophobia have fear-based reactions. Misophonia breeds aggression and phonophobia breeds anxiety. Although not widely considered by the medical community, some researchers have suggested that misophonia should be classified as a formal psychiatric disorder [53].

There is no consensus on the causes of hyperacusis, but some believe it is associated with tonic tensor tympani syndrome, a condition in which the tensor tympani muscle in the middle ear involuntarily contracts after a person is exposed to an intolerable sound exposure, thus reducing the threshold for activation of the acoustic reflex associated with the tensor tympani muscle [54]. Some researchers have claimed an association between the reaction of the auditory cortex of the brain to noise exposure and hyperacusis[55]. It has also linked with tinnitus, head injury, migraine, depression, autism, post-traumatic stress disorder, and an assortment of diseases, including Williams syndrome, Addison's disease, Bell's palsy, Lyme disease, perilymph fistula (mentioned earlier, in which the oval or round window ruptures and perilymph fluid leaks into the middle ear cavity), and Ramsay Hunt syndrome [56,57]. Physical discomfort usually accompanies moderate sound exposures for those experiencing this condition.

There have not been many research studies specific to the prevalence of hyperacusis, but one recent study in Brazil showed that 3% of elementary school children have this condition [58].

A significant problem with those suffering from hyperacusis or misophonia is that they are not diagnosed and treated for the disorder. There are treatments that have successfully reversed both of these conditions in many cases, so those having these symptoms should seek evaluations from an audiologist or otolaryngologist who works with these conditions.

3.4.2   Electrohypersensitivity

Although not directly related to sound, one documented symptom of hypersensitivity to electromagnetic fields (known as electrohypersensitivity or EHS) is hearing different types of sound that are associated with the electromagnetic fields (as discussed earlier). EHS is recognized as a functional impairment in Sweden, with roughly 3% of the population being affected [59]. Many studies have been performed to support EHS diagnoses [60] and groups have been organized to deal with these issues [61]. EHS is characterized by a host of symptoms including headache, dizziness, fatigue, concentration difficulties, memory loss, sleep disturbance, and skin sensations as the most commonly reported [62]. Complicating diagnoses associated with EHS are potentially confounding factors (unrelated to electromagnetic field exposures) and there have been research studies over the past 20 years that both support and dismiss the direct connection between electromagnetic field exposures and these symptoms [63]. These symptoms are also shared by those affected by allergies and chemical exposures, as well as those experiencing high levels of psychological stress. Independent of research study results, there is agreement that more research needs to be done to address this issue, as EHS has not been universally accepted as a diagnosable disease. Many people experiencing the types of hums discussed in Chapter 6 report similar symptoms to those suffering from what has been characterized as EHS.

3.5 Low-frequency and Infrasound Perception

One of the most controversial topics with regard to the effects of sound on people is how low-frequency sound and infrasound affect us. Low-frequency sound is generally considered as covering the range between 20 and 200 Hz, while infrasound is generally considered to represent frequencies below 20 Hz. With regard to auditory perception, Figure 3.3 shows the latest accepted hearing thresholds for low frequencies and infrasound to 4 Hz. Other earlier investigations [65,66] have resulted in similar results. A key point to note from the information in Figure 3.3 is that the general viewpoint that infrasound is not audible by humans is not correct; infrasound is just associated with higher thresholds than sounds in the 20 to 20,000 Hz range (typically considered as the audible range).

The studies resulting in the data shown in Figure 3.3 are based on recognition of pure tones. Studies using complex tones (combinations of two or more tones simultaneously,

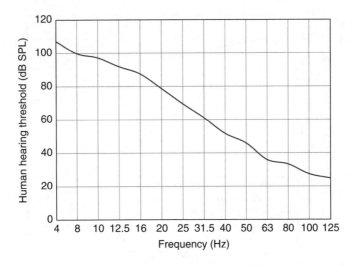

Figure 3.3 Mean threshold of human hearing from 4 to 125 Hz. SPL, sound pressure level. (Adapted from Watanabe and Møller [64], with permission of Multi-Science Publishing Co. Ltd.)

which are more in line with sounds we normally experience) have shown that thresholds up to 10 dB lower than those shown in Figure 3.3 can occur with many simultaneous tones of similar frequencies. These thresholds increased as the frequency separation between tones increased [67].

Hearing thresholds are generally the same for men and women until the age of 50 [68], after which the differences are noted in the discussion of presbycusis (loss of hearing with age) in Chapter 4.

Low-frequency sound and infrasound can excite resonances in the human body, which are sensed as vibration. Figures 3.4 and 3.5 show the thresholds of vibratory sensations in the body and head, respectively, with frequency. These figures cannot be combined to compare thresholds for the body and head as they cover different frequency ranges; however, the head thresholds are, in general, lower than the body thresholds. These figures show that the threshold for vibratory sensation is much higher than that for hearing. Sound-induced vibration response in the body, however, is not as consistent as the hearing response, because the vibration response is based on an individual's body mass index (BMI), defined as weight (in kg) divided by the square of the height (in m). Vibration responses generally decrease with increasing BMI for frequencies above 40 Hz in the chest area, while vibration responses to low-frequency sound decrease with increasing BMI for all frequencies in the abdomen. The chest area tends to induce higher vibration levels (by around 10 dB) than the abdomen for frequencies between 100 and 200 Hz [71]. The relationship to BMI is not necessarily related to vibrations below the surface in internal organs.

People who have significant deafness tend to sense low-frequency noise and infrasound mainly through the feelings of vibration induced by the signals, as the cochlea is not processing the signals for deaf people in the way they are processed for those with normal hearing. The threshold curves in Figures 3.4 and 3.5 therefore show the difference in

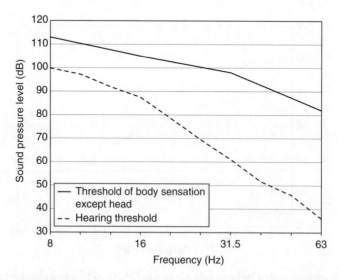

Figure 3.4 Threshold of body sensation and hearing from 8 to 63 Hz (adapted from Yamada [69] with permission of Multi-Science Publishing Co. Ltd.)

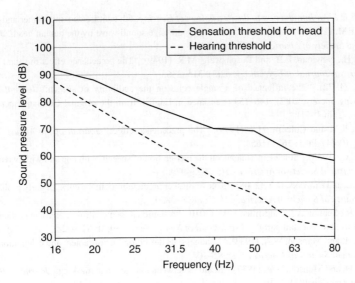

Figure 3.5   Threshold of head sensation and hearing from 16 to 80 Hz (adapted from Takahashi et al. [70] with permission of Multi-Science Publishing Co. Ltd.)

sensation thresholds for deaf people compared with those for people with normal hearing, with a difference of 10 dB or less for infrasound and 10–40 dB for low frequencies. These threshold differences for higher frequencies are in the 60 dB range.

References

[1] Fastl, H. and Zwicker, E. *Psychoacoustics: Facts and Models. Third Edition.* Heidelberg: Springer, 2007.

[2] Borg, E. and Counter, S A. (1989). "The middle-ear muscles." *Scientific American*, 261(2): 74–78.

[3] Fletcher, J.L. and Riopelle, A.J. (1960). "Protective effect of the acoustic reflex for impulsive noises." *Journal of the Acoustical Society of America*, 32(3): 401–404.

[4] Klockhoff, I. (1961). "Middle ear muscle reflexes in man." *Acta Oto-laryngologica*, Supplement 164.

[5] von Békésy, G. *Experiments in Hearing.* New York: McGraw-Hill Book Company, 1960.

[6] von Békésy, G. (1947). "The variation of phase along the basilar membrane with sinusoidal vibrations." *Journal of the Acoustical Society of America*, 19(3): 452–460.

[7] Cheatham, M.A. and Dallos, P. (2001). "Inner hair cell response patterns: implications for low-frequency hearing." *Journal of the Acoustical Society of America*, 110(4): 2034–2044.

[8] Yost, W.A. *Fundamentals of Hearing*, 5th edn. Bingley, UK: Emerald Group Publishing Limited, 2008

[9] Kemp, D.T. (1978). "Stimulated acoustic emissions from within the human auditory system." *Journal of the Acoustical Society of America*, 64(5): 1386–1391.

[10] Mukerji, S., Windsor, A.M. and Lee, D.J. (2010). "Auditory brainstem circuits that mediate the middle ear muscle reflex." *Trends in Amplification*, 14(3): 170–191.

[11] Horowitz, S.S. *The Universal Sense: How Hearing Shapes the Mind.* New York: Bloomsbury, 2012.

[12] Gardner, M.B. and Gardner, R.S. (1973). "Problem of localization in the median plane: effect of pinnae cavity occlusion." *Journal of the Acoustical Society of America*, 53(2): 400–408.

[13] Hebrank, J. and Wright, D. (1974). "Are two ears necessary for localization of sound sources on the median plane?" *Journal of the Acoustical Society of America*, 56(3): 935–938.

[14] Wright, D, Hebrank, J.H. and Wilson, B. (1974). "Pinna reflections as cues for localization." *Journal of the Acoustical Society of America*, 56(3): 957–962.

[15] Moore, B.C.J. *An Introduction to the Psychology of Hearing*, 6th edn. Leiden, the Netherlands: Brill, 2013.

[16] Wiener, F.M. (1947). "On the diffraction of a progressive sound wave by the human head." *Journal of the Acoustical Society of America*, 19(1): 143–146.

[17] Wallach, H., Newman, E.B. and Rosenzweig, M.R. (1949). "The precedence effect in sound localization." *The American Journal of Psychology*, 62(3): 315–336.

[18] Haas, H. (1972). "The influence of a single echo on the audibility of speech." *Journal of the Audio Engineering Society*, 20(2): 146–159 (translated in English from the original German paper published in *Acustica* 1(2) in 1951).

[19] Litovsky, R.Y. and Colburn, H.S. (1999). "The precedence effect." *Journal of the Acoustical Society of America*, 106(4), Pt. 1: 1633–1654.

[20] Cherry, E.C. (1953). "Some experiments on the recognition of speech, with one and with two ears." *Journal of the Acoustical Society of America*, 25(5): 975–979.

[21] Jones, G.L. and Litovsky, R.Y. (2008). "Role of masker predictability in the cocktail party problem." *Journal of the Acoustical Society of America*, 124(6): 3818–3830.

[22] Kerlin, J.R., Shahin, A.J. and Miller, L.M. (2010). "Attentional gain control of ongoing cortical speech representations in a "cocktail party"." *The Journal of Neuroscience*, 30(2): 620–628.

[23] Fletcher, H. and Munson, W.A. (1933). "Loudness, its definition, measurement and calculation." *Journal of the Acoustical Society of America*, 5(2): 82–108.

[24] Fletcher, H. and Munson, W.A. (1937). "Relation between loudness and masking." *Journal of the Acoustical Society of America*, 9(1): 1–10.

[25] Zwicker, E., Flottorp, G. and Stevens, S.S. (1957). "Critical band width in loudness summation." *Journal of the Acoustical Society of America*, 29(5): 548–557.

[26] Junqua, J. (1993). "The Lombard reflex and its role on human listeners and automatic speech recognizers." *Journal of the Acoustical Society of America*, 93(1): 510–524.

[27] Manabe, K., Sadr, E.I. and Dooling, R.J. (1998). "Control of vocal intensity in budgerigars (*Melopsittacus undulates*): Differential reinforcement of vocal intensity and the Lombard effect." *Journal of the Acoustical Society of America*, 103(2): 1190–1198.

[28] Golz, A., et al. (2001). "The effects of noise on the vestibular system." *American Journal of Otolaryngology*, 22(3): 190–196.

[29] Graybiel, A. and Knepton, J. (1976). "Sopite syndrome: A sometimes sole manifestation of motion sickness." *Aviation Space and Environmental Medicine*, 47(8): 873–882.

[30] Lawson, B.D. and Mead, A.M. (1998). "The Sopite syndrome revisited: Drowsiness and mood changes during real or apparent motion." *Acta Astronautica*, 43(3–6): 181–192.

[31] Kaski, D., et al. (2012). "The Tullio phenomenon: a neurologically neglected presentation." *Journal of Neurology*, 259(1): 4–21.

[32] Minor, L.B., et al. (1998). "Sound- and/or pressure-induced vertigo due to bone dehiscence of the superior semicircular canal." *Archives of Otolaryngology – Head & Neck Surgery*, 124(3): 249–258.

[33] Barr, T. (1886). "Enquiry into the effects of loud sounds upon the hearing of boilermakers and others who may work amid noisy surroundings." *Proceedings of the Philosophical Society of Glasgow*, 18: 223–239.

[34] Shimokura, R., et al. (2014). "Intelligibility of cartilage conduction speech in environmental noises." *Proceedings of ICBEN 2014*, Nara, Japan.

[35] Shimokura, R., et al. (2014). "Cartilage conduction hearing." *Journal of the Acoustical Society of America*, 135(4): 1959–1966.

[36] Rutka, J. *Hearing loss and tinnitus*. Toronto: Workplace Safety and Insurance Appeals Tribunal, 2013.

[37] Rauschecker, J.P., Leaver, A.M. and Mühlau, M. (2010). "Tuning out the noise: limbic-auditory interactions in tinnitus." *Neuron*, 66(6): 819–826.

[38] Roberts, L.E., et al. (2010). "Ringing ears: the neuroscience of tinnitus." *The Journal of Neuroscience*, 30(45): 14972–14979.

[39] Mahoney, C. J., et al. (2011). "Structural neuroanatomy of tinnitus and hyperacusis in semantic dementia." *Journal of Neurology, Neurosurgery & Psychiatry*, 82(11): 1274–1278.

[40] Henry, J.A., Dennis, K.C. and Schechter, M.A. (2005). "General review of tinnitus: prevalence, mechanisms, effects, and management." *Journal of Speech, Language, and Hearing Research*, 48: 1204–1235.

[41] Walford, R.E. (1983). "A classification of environmental "hums" and low frequency tinnitus." *Journal of Low Frequency Noise and Vibration*, 2(2): 60–84.

[42] Ritz, T., Adem, S. and Schulten, K. (2000). "A model for photoreceptor-based magnetoreception in birds." *Biophysical Journal*, 78(2): 707–718.

[43] Stevens, S.S. (1937). "On hearing by electrical stimulation." *Journal of the Acoustical Society of America*, 8: 191–195.

[44] Jones, R.C. Stevens, S.S. and Lurie, M.H. (1940). "Three mechanisms of hearing by electrical stimulation." *Journal of the Acoustical Society of America*, 12: 281–290.

[45] Perrott, D.R. and Higgins, P. (1973). "Notes on the electrophonic hearing effect." *Journal of the Acoustical Society of America*, 53(5): 1437–1438.

[46] Frey, A.H. (1962). "Human auditory system response to modulated electromagnetic energy." *Journal of Applied Physiology*, 17(4): 689–692.

[47] Frey, A.H. and Coren, E. (1979). "Holographic assessment of a hypothesized microwave hearing mechanism." *Science*, 206(4415): 232–234.

[48] Foster, K.R. and Finch, E.D. (1974). "Microware hearing: Evidence for thermoacoustic auditory stimulation by pulsed microwaves." *Science*, 185, pp. 256–258.

[49] Lin, J.C. (2001). "Hearing microwaves: The microwave auditory phenomenon." *IEEE Antennas and Propagation Magazine*, 43(6): 166–168.

[50] Lin, J.C. and Wang, Z. (2007). "Hearing of microwave pulses by humans and animals: effects, mechanism, and thresholds." *Health Physics*, 92(6): 621–628.

[51] Elder, J.A. and Chou, C.K. (2003). "Auditory response to pulsed radiofrequency energy." *Bioelectromagnetics*, 24(S6): S162–S173.

[52] Cassie, D. (2012). "Misophonia: Beyond irritation to a hatred and aversion of sound: A small group of audiologists are looking into this curious area." *The Hearing Review*, 19(5):52–53.

[53] Schröder, A., Vulink, N. and Denys, D. (2013). "Misophonia: Diagnostic criteria for a new psychiatric disorder." *PLOS ONE*, 8(1): 1–5.

[54] Westcott, M. (2010). "Hyperacusis: A clinical perspective on understanding and management." *New Zealand Medical Journal*, 123(1311): 154–160.

[55] Sun, W., et al. (2012). "Noise exposure enhances auditory cortex responses related to hyperacusis behavior." *Brain research*, 1485: 108–116.

[56] Schwartz, P., Leyendecker, J. and Conlon, M. (2011). "Hyperacusis and misophonia: The lesser-known siblings of tinnitus." *Minnesota Medicine*, 94(11): 42–43.

[57] Baguley, D.M. (2003). "Hyperacusis." *Journal of the Royal Society of Medicine*, 96(12): 582–585.

[58] Coelho, C.B., Sanchez, T.G. and Tyler, R.S. (2007). "Hyperacusis, sound annoyance, and loudness hypersensitivity in children." *Progress in Brain Research*, 166: 169–178.

[59] Johansson, O. (2010). "Aspects of studies on the functional impairment electrohypersensitivity." *Earth and Environmental Science*, 10(1): 1–7.

[60] Rea, W.J., et al. (1991). "Electromagnetic field sensitivity." *Journal of Bioelectricity*, 10(1–2): 241–256.

[61] Grant, L. *The Electrical Sensitivity Handbook*. Prescott, Arizona: Weldon Publishing, 1995.

[62] Seitz, H., et al. (2005). "Electromagnetic hypersensitivity (EHS) and subjective health complaints associated with electromagnetic fields of mobile phone communication – a literature review published between 2000 and 2004." *Science of the Total Environment*, 349: 45–55.

[63] Rubin, G.J., Munshi, J.D. and Wessely, S. (2005). "Electromagnetic hypersensitivity: A systematic review of provocation studies." *Psychosomatic Medicine*, 67(2): 224–232.

[64] Watanabe, T. and Møller, H. (1991). "Low frequency hearing thresholds in pressure field and free field." *Journal of Low Frequency Noise and Vibration*, 9(3): 106–115.

[65] Yeowart, N.S., Bryan, M.E. and Tempest, W. (1967). "The monaural M.A.P. threshold of hearing at frequencies from 1.5 to 100 c/s." *Journal of Sound and Vibration*, 6(3): 335–342.

[66] Yeowart, N.S. and Evans, M.J. (1974). "Thresholds of audibility for very low-frequency pure tones." *Journal of the Acoustical Society of America*, 55(4): 814–818.

[67] Ryu, J., et al. (2011). "Hearing thresholds for low-frequency complex tones of less than 150 Hz." *Journal of Low Frequency Noise, Vibration and Active Control*, 30(1): 21–30.

[68] Møller, H. and Pedersen, C.S. (2004). "Hearing at low and infrasonic frequencies." *Noise & Health*, 6(23): 37–57.

[69] Yamada, S. et al. (1983). "Body sensations of low frequency noise of ordinary persons and profoundly deaf persons." *Journal of Low Frequency Noise and Vibration*, 2(3): 32–36.

[70] Takahashi, Y. (2013). "Vibratory sensation induced by low-frequency noise: The threshold for 'vibration perceived in the head' in normal-hearing subjects." *Journal of Low Frequency Noise, Vibration and Active Control*, 32(1+2): 1–9.

[71] Takahashi, Y., et al. (1999). "A pilot study on the human body vibration induced by low frequency noise." *Industrial Health*, 37:28–35.

4

Physiological Effects of Sound Exposure

4.1 Introduction

Although there are many physiological effects of sound on people included in the published literature, there is universal agreement on only one of those effects – noise-induced hearing loss (NIHL). Many research studies have linked other physiological effects to noise, but the effects that have the most research support are related to cardiovascular diseases. Isolated research has recently unveiled potential new noise-related illnesses related with low-frequency noise and infrasound exposure, most notably vibroacoustic disease, but these links have not as yet been accepted by the overall medical community. This chapter summarizes the most recent research in these areas.

4.2 Body Resonance and Damage Potential

Chapter 3 (Section 3.5) touched on the subject of body resonance in the discussion related to thresholds of sound associated with body and head sensations. Much of this is associated with resonance of different parts of the human body. The concept of resonance was discussed in Chapter 1, related to room modes and standing waves, phenomena associated with air resonance. Body resonance is associated with mechanical resonance, a property of any solid material that causes it to vibrate at a sympathetic, higher magnitude near a specific frequency than it would when that material is exposed to sound in any other frequency range.

Figure 4.1 illustrates the general principles behind mechanical resonance of a homogeneous material. Every material object has a resonance or natural frequency associated with it, which is a function of its density (mass per unit volume). When an object is exposed to sound dominated by energy at its resonance frequency, it will vibrate at an amplified level.

The Effects of Sound on People, First Edition. James P. Cowan.

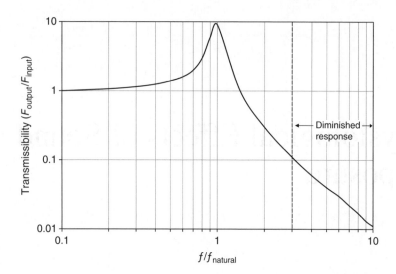

Figure 4.1   Generic resonance response of a homogeneous material on a logarithmic scale

Vibrational energy produces forces on a material. The force transmissibility (output/input) shows amplification (when transmissibility is greater than 1) at the natural frequency and significant attenuation (output being less than the input force) when the forcing frequency is more than three times the natural frequency. If the vertical scale in Figure 4.1 is converted to decibels (dB), each factor of 10 would translate to a change of 10 dB (e.g., a transmissibility of 1 would translate to 0 dB, a transmissibility of 10 would translate to a 10 dB amplification, and a transmissibility of 0.1 would translate to an attenuation of 10 dB). The amount of amplification at the natural frequency depends on the damping or stiffness of the material. A high amount of damping (vibration suppression) would lower the amplification and a low amount of damping would raise it.

The large variability of densities in the many components of the human body makes it difficult to determine resonance frequencies of individual parts. Resonance frequencies of different parts of the human body have been studied extensively, with a consensus that the human body in general has fundamental resonance frequencies in the range of 5–12 Hz and individual components have resonant frequencies up to 60 Hz [1]. Skeletal resonance frequencies are higher, with the skull resonating at a fundamental frequency in the 900–1,200 Hz range [2]. Resonance phenomena can result in discomfort from tonal exposures at high levels (above 120 dB), but most natural and man-made sounds do not approach levels that could cause any body resonance-related issues for people.

Thresholds of aural pain occur at extremely high exposure levels, in the 120 dB range for mid-frequencies (500–1,000 Hz), increasing with decreasing frequency to 145 dB at 20 Hz and 165 dB at 2 Hz. Thresholds for damage to human organs (other than the inner ear hearing organs) are of the order of 175–180 dB through the mid-frequency range, at which point the mechanical limits of the eardrum and middle ear ossicles are surpassed. This can cause eardrum rupture or dislodging of the ossicles. Lung damage has also been reported at these sound pressure levels [1].

Respiratory rhythm issues, nausea, headaches, visual blurring, choking, and fatigue have been reported for exposures of 150 dB at 50–100 Hz [3]. Uncontrollable eye movements, known as nystagmus, and vestibular responses have also been reported beginning at levels of 135 dB in the 1,000–1,500 Hz range, and at 145–160 dB for 2,000 Hz. All of these exposure limits are independent of exposure durations. For frequencies below 100 Hz, nystagmus limits vary according to exposure time, for which they level off at 60 seconds. For a 1-minute exposure, the threshold is 110 dB at 20 Hz, 120 dB at 10 Hz, and 130 dB at 5 Hz. These limits increase dramatically as exposure times decrease [1]. In no cases have exposures less than 110 dB been reported to cause impairing physiological effects. Physical damage to any body tissue has only been reported for exposures above 170 dB.

4.3 Hearing Loss

Hearing loss is manifested in many forms and has many potential causes. The most preventable form of hearing loss is caused by high noise exposures. In any case, hearing loss can have life-altering effects, as hearing plays a major role in communications and social interactions. The suffix "acusis" relates to hearing (from the Greek word *akousis*), and most often to hearing loss. The most common causes of hearing loss are the aging process (known as presbyacusis or presbycusis), high noise exposure (known as socioacusis or sociocusis for non-occupational exposures), medical conditions (sometimes referred to as nosoacusis or nosocusis), and medications. Some chemical exposures can also cause hearing loss.

There are generally two categories of hearing loss, related to the locations in the hearing mechanism in which the causes occur. These are conductive and sensorineural. Conductive hearing loss is associated with issues before sound waves enter the inner ear (usually in the middle ear) and sensorineural hearing loss is associated with the neural system between the cochlea and the brain. Each type of hearing loss has a characteristic frequency response that is discussed in the following.

4.3.1 Presbycusis

Presbycusis is hearing loss associated with the aging process. It is usually sensorineural, although some age-related hearing loss can occur from diminished responses of the middle ear ossicles. Presbycusis usually affects frequency discrimination above 1,000 Hz, as shown in Figure 4.2 from measured data and Figure 4.3 from the ISO standard for threshold corrections due to age. As the figures show, males tend to be more affected by presbycusis than females; however, it is assumed that men are exposed to higher noise levels than women, so these differences may not be as dramatic as shown in these figures if noise exposures had been taken into account. This frequency response trend is independent of exposure to other hearing loss agents, such as medical conditions and medications. Outer hair cell loss, loss of afferent neurons in the cochlea, and atrophy of the stria vascularis are contributing factors to presbycusis [4].

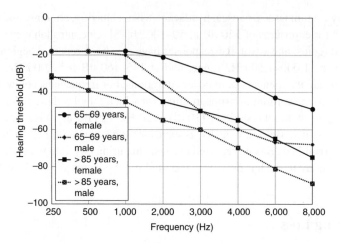

Figure 4.2 Typical hearing thresholds for elderly females and males (age in years) (adapted from data in Schmiedt [4], Dubno et al., [5] and Lee et al. [6])

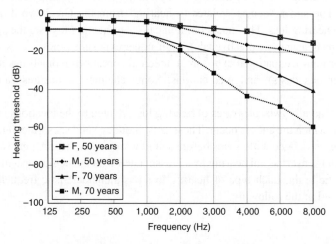

Figure 4.3 Median hearing thresholds for females (F) and males (M) at 50 and 70 years of age (adapted from data in ISO 7029-2000 [7])

It is mostly thought that presbycusis is caused by the gradual deterioration of the delicate neural system associated with hearing as one ages. An interesting counter-argument to that mode of thought comes from a study that was performed in the 1960s on an isolated tribe in the Sudan that was not exposed to the noises of most of contemporary society [8]. These people exhibited slower signs of presbycusis than is typically seen, raising the possibility that presbycusis is related not only to the aging process but also to typical non-occupational environmental noise exposures over time (known as sociocusis), although stress and diet may affect these results as well.

In addition to hearing tests, blood pressures of the tribal members were tested and it was noted that, not only did blood pressure remain essentially unchanged during the entire life span (from ages 10 to 90), but comparisons were made with American blood pressure data at the time. Mean systolic blood pressures of 75-year-olds were 115 mm of mercury (mmHg) for tribe members compared with more than 145 mmHg for Americans. One of many potential causes of presbycusis is the deterioration of small blood vessels in the cochlea caused by high blood pressure and atherosclerosis, both of which conditions these tribe members were lacking. This stresses the importance of including all potential causative factors in any of these types of studies.

In any case, it is difficult to separate pure presbycusis from hearing loss caused by other sources, but the spectral trends associated with presbycusis are universally accepted, that being minimal hearing loss below 1,000 Hz and hearing loss increasing with increasing frequency above 1,000 Hz. The focus of information from Figures 4.2 and 4.3 should therefore be more the spectral trend than the absolute threshold values. Presbycusis has been documented to begin as early as the teenage years [9], but it becomes most pronounced after age 50.

It is worth noting that hearing thresholds below 1,000 Hz, and especially in the low-frequency range below 200 Hz, are not significantly affected by age [10]. Therefore, low-frequency noise can generate annoyance issues at any age, supporting the more current arguments against low-frequency noise and their potential effects on people.

4.3.2   Noise-induced Hearing Loss

Published research linking noise exposure with hearing loss has a long history, most notably beginning with the introduction of deafness associated with boilermakers and blacksmiths in the 19th century [11–13]. Before 20th-century laws were instituted to protect workers from occupational hearing loss, it was merely accepted that workers in certain industries associated with high noise would lose most of their hearing abilities. However, this hearing loss was reserved for the minority of the population working in these especially noisy occupations. Contemporary society exposes nearly everyone to high levels of noise, not at all limited to the occupational setting. NIHL is the leading occupational disease and, according to the World Health Organization, is "the most prevalent irreversible occupational hazard [14]." In addition to that, non-occupational NIHL is also prevalent due to recreational activities involving high noise levels such as firearm activities, power tools, and entertainment facilities (involving live and recorded sources).

Noise-induced hearing loss is associated with the weakening and deterioration of the hair cells in the inner ear. The outer hair cells are more susceptible to damage from high noise levels than the inner hair cells as they are in direct contact with the tectorial membrane and therefore they are exposed to excessive shearing motions in the organ of Corti, although there have been reports that the tips of inner hair cells may be in contact with the tectorial membrane [15]. There have also been lesions found in the stria vascularis resulting from tears in Reissner's membrane, which regulates the mixing of fluids in the cochlea [16]. The death of hair cells is more extensive in the basal end of the cochlea (closer to the oval window) than at the apical end (closer to the helicotrema), eventually causing extensive degeneration of the auditory nerve [17].

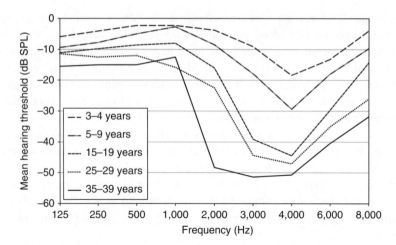

Figure 4.4 Hearing thresholds as a function of years working in a facility with an 8-hour average exposure level of 100 dBA (adapted from Taylor, W., et al., [18] with permission of the Acoustical Society of America)

This trends to higher-frequency losses as the basal hair cells send high-frequency information to the brain and the apical hair cells send lower-frequency signals to the brain. A combination of this with human hearing sensitivity maximizing in the 3,000–4,000 Hz range due to ear canal amplification results in NIHL being most prevalent in the 3,000–4,000 Hz range.

Figure 4.4 shows the typical hearing threshold trend of NIHL. These data were compiled from a study of female jute weavers in Scotland in the 1960s [18]. As occupational noise regulations were instituted worldwide in the 1970s, it would be impossible to have such a pure study of NIHL since that time. A perfect laboratory for this study was provided by the jute weaving industry in Scotland, in which employees had spent up to 50 years continuously working in the same building, if not the same weaving loom, where background sound levels averaged in the 99–102 dBA range at the operators' positions during all working hours. Using females in the study also lowered the potential for confounding factors as they tended to not engage in high-noise recreational activities such as those the male population were engaged in. The subjects were also screened for medical conditions that were known to lead to hearing loss.

As for presbycusis, NIHL does not significantly affect the perception of frequencies below 1,000 Hz. Differing from presbycusis is the characteristic dip or notch in the 3,000–4,000 Hz range of the hearing threshold curve. The first sign of NIHL is a dip in hearing threshold in the 4,000 Hz range. As exposure times increase, this dip broadens to 2,000 Hz as thresholds progressively increase. Since there is a wide range in hearing threshold changes from noise exposures, these trends are illustrated here in terms of median (50% of the population) values. ISO Standard 1999:2013 provides equations for calculating these thresholds for 10%, 50%, and 90% of the population. Figure 4.5 shows these results for the median (50%) values related to 8-hour daily exposures to 90 and 100 dBA for 10 and 40 years.

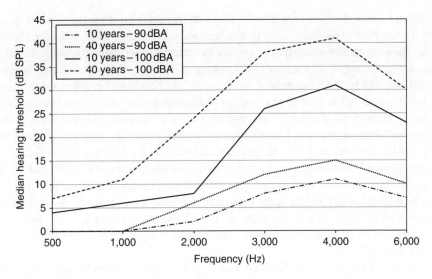

Figure 4.5 Median hearing thresholds as a function of years for 8-hour average exposure levels of 90 and 100 dBA (adapted from data in ISO 1999:2013 [19])

Unless caused by traumatically high sound level exposure, NIHL occurs gradually over time, with initial exposures causing a temporary threshold shift (TTS), which, after a period away from the high noise levels, recovers to normal or improved hearing conditions. Extended exposures to high noise levels results in a permanent threshold shift (PTS), which is the basis for the data illustrated in Figures 4.4 and 4.5, but there is no clear correlation for when a TTS becomes a PTS [20]. Although TTS has been considered to be harmless in the past, current research has shown that exposures causing TTS can cause permanent cochlear damage [21]. It is generally recognized that an average occupational exposure less than 90 dBA for 40 hours/week will not cause NIHL for 85% of the population and exposures less than 85 dBA under the same conditions are safe for 95% of the population [22]. This assumes a minimum of a 10-year exposure of 40 hours/week for any significant effects. It is also agreed that acceptable exposure times should be cut in half with each doubling of sound energy above the 8-hour exposure limit, up to a maximum permitted exposure of 115 dBA independent of exposure time. This is the basis for occupational hearing conservation legislation, which is discussed further in Chapter 8.

The question of susceptibility to NIHL has been the subject of much research. Although results reported herein are based on either average or median thresholds, there is a large amount of variability in these values between individuals. The reasons for this are unknown, but studies have been performed related to the effects of such non-auditory factors as eye color, gender, age, smoking, and genetics.

With regard to eye color, the pigmentation of the eye is associated with melanin, which is also present in the many of the components of the inner ear. Several studies have supported the premise that people with brown eyes are less susceptible to NIHL than those with blue eyes, by a factor of 2–3 dB, especially in the critical speech intelligibility range of 3,000–6,000 Hz [23].

Gender and age studies were discussed earlier with regard to presbycusis, showing in most cases women being less susceptible to hearing loss with age than men; however, there are studies showing minimal difference between the genders, suggesting the published differences are attributable to sociocusis (assuming men are exposed to higher levels of occupational and recreational noise than women) [24]. Smoking can affect hearing acuity because it introduces several agents that can be harmful to the hearing mechanism, known as ototoxic agents, most notably carbon monoxide. Carbon monoxide exposure by itself has not been proven to cause hearing loss, but carbon monoxide during high noise level exposures has been reported to enhance inner ear hair cell loss, resulting in more permanent hearing threshold shifts than would result from the noise exposure alone [25]. As for NIHL alone, the outer hair cells seem to be more vulnerable to enhanced damage than inner hair cells. As mentioned in Chapter 3, the inner hair cells have many more afferent neural connections with the brain than do the outer hair cells; however, the outer hair cells are thought to amplify and enhance acoustic signals that can play a significant role in hearing loss if they are damaged.

There have been several studies addressing the potential for lowering the risk of NIHL by introducing a lower level sound to "condition" the hearing mechanism into a protective state. The protective role of the middle ear muscles, which can reduce noise exposure to the inner ear by as much as 30 dB after a 100 ms delay, has been discussed in Chapter 3. There appear to be other protective mechanisms in addition to the acoustic reflexes associated with these muscles, such as changes in the cochlea and the efferent (path from the brain to the cochlea) auditory nervous system, which are activated when moderate noise exposures are presented before a high-level exposure [26]. This moderate level is said to condition or toughen the system to reduce the negative effects of noise exposure on the hearing mechanism.

Conditioning tests have consistently shown the effectiveness of moderate (around 80 dB) pre-exposures of noise in reducing the damaging effects of subsequent high levels of noise exposure by up to 20 dB [27,28]. Studies have also proved that the middle ear muscles are not involved in this conditioning effect [29]. These conditioning sounds must be in a similar frequency range to the higher-level noise to be effective. For example, a low-frequency conditioning signal is not effective in protecting against NIHL from a high-frequency signal. Experiments have actually shown that this situation makes the auditory system more vulnerable to threshold shifts from high-level noise exposures [30].

Another aspect of the relationship between NIHL and noise exposures is that interrupted exposures produce less TTS and PTS than continuous exposures of equal acoustic energy [31]. Continuous exposures reach a limit of threshold shift after 8–24 hours, known as the asymptotic threshold shift, which has been shown to be stable for up to several years of continuous exposure. Intermittent exposures of equal energy to continuous exposures can produce a threshold shift of up to 30 dB less.

Children are the most vulnerable group to NIHL, because their hearing mechanisms are developing and they are mostly unaware of the potential hazards of high-level noise exposures. In the United States alone, it was estimated in 2001 that 5.2 million children between the ages of 6 and 19 years had a noise-induced threshold shift, the majority of which were in one ear [32]. In an updated version of this survey, published in 2010, a significant increase in noise-induced threshold shifts among female youths was reported, with an overall increase in recreational noise exposure [33].

A recently described variation of NIHL is known as acoustic shock injury (ASI) [34]. Originally associated with call centers in Australia, ASI was associated with a sudden, unexpected, high-level sound randomly transmitted through a telephone line which causes pain in and around the ear, a feeling of aural fullness, distorted hearing, headache, tinnitus, hyperacusis, and vertigo. Hearing loss is not always a symptom of ASI and symptoms can last from hours or days to an indefinite time period. ASI is thought to be caused by tonic tensor tympani syndrome, a condition in which the tensor tympani muscle in the middle ear becomes uncontrollably active. Electronic and behavioral therapies can be effective for treating this malady.

An agent that is important in protecting the inner ear from noise-induced damage is magnesium. Experiments have shown PTS differences of up to 20 dB from the same noise exposure when comparing magnesium concentrations in the perilymph fluids of the inner ear [35]. Other promising research in this field has revealed that NIHL is caused not only by physical injury to the hearing organs but also by oxidative stress on those organs. This has led to the conclusion that some NIHL can be prevented or treated with antioxidants [36,37].

Until recently, genetics was not considered a real factor in NIHL susceptibility. Genetic research is now revealing links that will result in effective prevention and treatment strategies for NIHL [21]. It has been accepted until recently that PTS associated with NIHL is irreversible, but genetic and stem cell research is progressing to the point where the regeneration of hair cells will be possible. This exciting research will provide treatments for NIHL that have not been possible in the past. Humans cannot regenerate cochlear hair cells but non-mammals (such as birds and cold-blooded vertebrates) can [38]. Ongoing studies to determine why some species can regenerate hair cells and others cannot will provide ground-breaking information for curing NIHL in people.

4.3.3 Hearing Loss from Illness or Agents

Although this book is focusing on the effects of sound on people, a summary of our loss of sound perception from other agents is worth considering here. Noise is not the only potential cause of hearing loss. Besides aging, other common causes include [39]:

- ear canal wax impaction
- otosclerosis
- otitis media (middle ear infection)
- head injury
- Ménière's disease
- acoustic neuroma (a benign tumor growing around the auditory vestibular nerve)
- perforated eardrum
- other infections
- genetic disorders
- ototoxic chemicals and medications.

These provide confounding factors when attempting to link hearing loss to a specific cause, especially noise.

Although it has been known for some time that some medications, especially certain antibiotics, can cause hearing loss, little has been known or published until recently regarding the potential effects of chemicals on hearing. These chemical exposures can take place in both occupational and environmental settings. The effects of ototoxic chemicals are not limited to the cochlea, as these agents are carried through the bloodstream to potentially affect all organs involved with hearing. Testing has shown that the central auditory pathway, including the seventh cranial nerve (known as the facial nerve, associated with facial muscles, including the muscles associated with acoustic reflex), can be affected by exposure to these agents [40]. The facial nerve is located adjacent to the auditory vestibular nerve in the area shared by the two cranial nerves.

Initial hearing threshold deficiencies associated with ototoxic chemicals are similar to those for NIHL, with a dip in the 3,000–6,000 Hz range, so these exposures should be addressed when evaluating anyone for NIHL. Extended exposures have been shown to affect hearing thresholds below 3,000 Hz [41], offering an opportunity to diagnose NIHL separately from chemical-induced hearing loss (or hearing loss from a combination of noise and chemical exposures). Some agents cause hearing loss by themselves, but most augment NIHL when noise exposures are involved and have not shown evidence of causing hearing loss except at high levels.

The general categories of chemicals reported to be involved with hearing loss are solvents, metals, asphyxiants, and pesticides [42]. Specific agents that have been most documented to increase hearing loss in the presence of noise include the following [43]:

- Trichloroethylene (primarily used as a grease remover but also found in some dry cleaning agents, carpet cleaners, paints, waxes, pesticides, adhesives, and lubricants)
- Xylene (found in paints, varnishes, and thinners)
- Styrene (used in producing plastics, synthetic rubber, resins, and insulation materials)
- Carbon disulfide (used in solvents and insecticides)
- Toluene (found in motor vehicle emissions, paints, lacquers, adhesives, rubber, and leather tanning)
- Carbon monoxide (mentioned earlier)
- Metals, such as arsenic, mercury, tin, manganese, and lead.

Except under high-level exposures, a single daily exposure to these agents has not been proven to be associated with hearing loss [44]. Studies linking these exposures with hearing loss reference exposures of at least 5 years [45]. Anyone interested in pursuing this subject further can use the references at the end of this chapter as a starting point.

4.4 Cardiovascular Disease

Until recently, the only published physiological effect related to noise, except at extremely high-level exposures (over 150 dB), was NIHL. Stress-related illnesses were casually mentioned but comprehensive research studies had rarely been performed. The psychological stress caused by noise can affect the endocrine system to release stress hormones that can affect the cardiovascular system [46]. There is now a sizable body of research attempting to link noise exposures with cardiovascular diseases. The biggest challenge associated with these studies is isolating this link from the many potential confounding factors that are also involved in causing cardiovascular disease.

Most of the recent studies attempting to prove this association are related to transportation noise (mainly highway and aircraft, with some isolated rail studies). Highway noise studies are more related to a steady diurnal noise exposure, while aircraft and rail noise studies are mostly related to night-time disturbance (especially awakenings caused by discrete pass-by events).

Studies relating cardiovascular disease and noise exposure are mostly divided into two subject categories – hypertension (high blood pressure) and ischemic diseases (blood flow restrictions). Although it can be argued that the link between cardiovascular disease and noise is stress-related and therefore psychologically (rather than physiologically) based, this discussion is being included with physiological effects, as it is in most of the literature.

A measure of a stress-related reaction is an evaluation of hormones released by the endocrine system, most notably the release of adrenaline and noradrenaline from a fight-or-flight reaction or cortisol from a defeat reaction [47]. The personal interpretation of the sound is also a component, so the sound level is not always an indicator of these effects. This is discussed further in Chapter 5.

Most of the studies linking noise with cardiovascular diseases are based on statistical analyses. The odds ratio (OR) is often used to quantify this link, and care must be taken to ensure that the results are interpreted correctly. The OR defines the chance of an effect being linked to a specific cause, or the chance of an effect occurring in one group divided by the chance of it occurring in another. An OR of 1 means there is no link between the stated cause and effect. An OR > 1 implies a potential link, with higher values implying a stronger potential link. Normally tied to an OR is a 95% confidence interval (CI), which defines the statistical significance of the results and is usually listed as a range of values. If that range has the value of 1 within it, the cause–effect link is not statistically significant. Another descriptor often used in these analyses is the relative risk (RR), which is based on probabilities of results occurring rather than the specific occurrences used in the calculation of OR. Although OR and RR can be significantly different for higher values, they are similar for relatively small risks.

An example can be helpful to explain this. There are four parameters that need to be established to perform these calculations:

- The probability of an agent causing an effect, a
- The probability of an agent not causing an effect, b
- The probability of producing an effect without exposure to the agent, c
- The probability of not producing an effect without exposure to the agent, d.

If these values are all percentages, then $b = a - 100$ and $d = c - 100$. The definition of the RR is:

$$\text{RR} = \frac{a/(a+b)}{c/(c+d)} \tag{4.1}$$

If a is much smaller than b, and c is much smaller than d, equation (4.1) reduces to the definition of the OR, which is:

$$OR = \frac{a/b}{c/d} = \frac{ad}{cb} \tag{4.2}$$

Assuming a hypothetical case of the cause-and-effect relationship between mold exposure and allergic reactions, say 50% of the exposed population develops allergic reactions to mold, while 2% of the surveyed population not exposed to mold develop the same allergic reactions. In this case, $a = 50$, $b = 50$, $c = 2$, and $d = 98$, and RR would be 25 and OR 49. The OR is significantly higher than the RR, which can result in different interpretations. However, if 5% of the exposed population develop these allergic reactions while 4% of the unexposed population develop the same allergic reactions, then $a = 5$, $b = 95$, $c = 4$, $d = 96$, RR = 1.25, and OR = 1.26. RR and OR are therefore similar for the case of a slight link between the agent and the effect. Because most of the OR and RR values associated with the studies listed in the following sections are in the 1.0–2.0 range, OR and RR values can be assumed to be similar for these purposes.

4.4.1 Hypertension

Hypertension is typically classified as a systolic blood pressure ≥ 140 mmHg or a diastolic blood pressure ≥ 90 mmHg. There have been several recent studies evaluating the link between hypertension and noise. These studies are mostly related to transportation noise, but some are related to general occupational exposures. In one study performed over a 2- to 4-year period, both positive and negative effects were identified, depending on the complexity of the tasks being performed [48]. Systolic blood pressure was shown to increase by 6% for workers with high job complexity and high (>80 dBA) noise exposure while a 4% increase in systolic blood pressure was found for workers in low noise environments and low job complexity. There was also no increase in systolic blood pressure for workers with high job complexity and low noise level exposure. Therefore, the combination of task complexity and noise level exposure can be a contributor to changes in blood pressure.

A large multinational study on the relationship between hypertension and noise exposure known as the HYENA study (Hypertension and Exposure to Noise near Airports) was performed between 2005 and 2006 [49]. For this study, nearly 5,000 people (split nearly evenly between men and women) between the ages of 45 and 70 living (for at least 5 years) near one of six major European airports in the UK (London Heathrow), Germany (Berlin Tegel), the Netherlands (Amsterdam Schiphol), Sweden (Stockholm Arlanda), Italy (Milan Malpensa), and Greece (Athens Elephterios Venizelos) were surveyed. The noise sources evaluated for this study were aircraft and road traffic noise. Noise exposures in terms of 24-hour equivalent sound level (L_{eq}) were evaluated for road traffic noise between 45 and 70 dBA. Aircraft noise exposures were evaluated in terms of 16-hour L_{eq} (between 7:00 am and 11:00 pm or between 6:00 am and 10:00 pm) between 35 and 70 dBA and L_{night} (8-hour L_{eq} between 11:00 pm and 7:00 am or between 10:00 pm and 6:00 am) between 30 and 60 dBA. Confounding factors included in the analyses were country, sex, age, alcohol consumption, body mass index, physical activity, and education. Smoking was explicitly not included in the model because the authors of the study stated that smoking is not a risk factor for hypertension, although the project report acknowledges that blood pressure increases after smoking and so study participants were required to refrain from smoking at least 30 minutes prior to any blood pressure testing.

The results of the HYENA study found that there are links between hypertension and road traffic noise at all times and aircraft noise for night-time operations. More specifically, a 10 dBA increase in L_{night} for aircraft noise was associated with an OR of 1.14 (and 95% CI of 1.01–1.29) for a hypertension link, and road traffic noise 24-hour L_{eq} exposures > 65 dBA were associated with an OR of 1.54 (and 95% CI of 0.99–2.40). Other results of this study showed that, for 24-hour L_{eq} roadway traffic noise values between 45 and 65 dBA, independent of the actual level, residency of more than 25 years was linked with hypertension (OR = 1.21, 95% CI: 1.05–1.41). Along the same lines, those who kept living room windows open in those noise environments were linked with hypertension (OR = 1.23, 95% CI: 1.07–1.43) [50].

A 2012 summary and analysis of 24 studies performed between 1970 and 2010 on the association between hypertension and road traffic noise exposures concluded that there was a slight link between the two for 16-hour L_{eq} ranging from 45 to 75 dBA [51]. This conclusion is based on an OR of 1.034 (and 95% CI of 1.011–1.056) for each increase of 5 dBA in 16-hour L_{eq}.

A recent Swiss study of the link between hypertension and rail and roadway traffic noise only suggested correlations for those with chronic illnesses such as diabetes [52].

Although most of these studies deal solely with the effects of noise on the adult population, there have been some studies attempting to link hypertension to noise among children. A recent study in Belgrade, Serbia, addressed this subject for school children aged 7–11 in six primary schools [53]. Basing its results on categorizing the difference between noisy and quiet daytime school exterior environments by whether the daytime L_{eq} was greater than or less than 60 dBA, and between noisy and quiet night-time home exterior environments by whether the night-time L_{eq} was greater than or less than 45 dBA, the study concluded that children in noisy schools (independent of the home noise environment) had higher systolic blood pressures (by 4–9 mmHg) than those in quiet schools. These values accounted for the confounding factors of gender, age, body mass index, family history of hypertension, and family income.

Yet another confounding factor that is difficult to separate from these analyses is air pollution, as most studies linking noise with hypertension are related to roadway traffic noise and residences close enough to roadways to be included in these studies are also exposed to significant air pollution.

The general consensus among research in this area is that there is a positive relationship between transportation noise exposure and hypertension, but the disagreement lies in the degree to which that association can be made. There is not enough information to support a clear dose–response relationship between hypertension and noise, but studies generally present increased risks for hypertension when roadway traffic noise levels exceed 65 dBA during the day and 55 dBA at night. That risk also exists for 10 dBA increases in average aircraft noise levels in the range of 45–70 dBA [54].

4.4.2   Ischemic Diseases

Studies attempting to link ischemic heart diseases to noise exposure mostly address myocardial infarction (MI) or heart attack, while some include links to strokes. Before the year 2000, few studies supported the link between environmental noise exposures and ischemic heart diseases [55]. Since that time, summaries of multiple studies have been published reporting slight but clear links noted between environmental noise exposures and ischemic heart diseases [56]. These studies are mostly from European countries and only deal with adults, with the highest association being for men exposed to average daily highway traffic noise levels exceeding 65–70 dBA for at least 10 years. For one study using data from 32 hospitals in Germany for people 20–69 years of age, the OR was 1.8 (95% CI: 1.0–3.2), considering such confounding factors as family history of MI, smoking, educational level, marital status, employment status, body mass index, hypertension, and diabetes [57]. Traffic noise levels below 70 dBA had mixed results.

A recent Danish study based on more than 50,000 participants showed a 10% (95% CI: 1.0–1.2) increase in risk of MI with each 10 dBA increase in average roadway traffic noise levels between L_{den} values of 42 and 84 dBA [58]. Confounding factors considered in this study included age, gender, smoking status, smoking duration and intensity, diet (specifically, intake of fruits, vegetables, and alcohol), body mass index, physical activity, education, diabetes, cholesterol, hypertension, and air pollution. This is one of the few studies that includes adjustments to eliminate traffic-related air pollution as a confounding factor, mainly in terms of nitrogen oxide (NO_x) concentrations. This study was also valuable in that it only addressed first cases of MI, thereby eliminating preventive medications from the mix. Another point to bear in mind about this study is that the risk results are in terms of an incidence rate ratio (IRR) rather than the OR that has been referenced in other studies. When the effects are relatively small, the IRR is comparable to the OR, which is the case for these studies.

A recent Canadian study of over 400,000 people has published results that also support the elimination of air pollution as a confounding factor in the link between roadway traffic noise and heart disease. This was based on a 5-year exposure of 45- to 85-year-olds to L_{den} values of 58–70 dBA [59]. This study showed a 22% increase in coronary heart disease mortality for L_{den} exposures > 70 dBA compared with that for people exposed to L_{den} values < 59 dBA. Comparing results for the same groups, there was a 4% increase in coronary heart disease mortality from the associated elevation in black carbon air pollution levels.

Enough data have been compiled for dose–response relationships to be developed between continuous environmental noise exposure and ischemic heart diseases. Figure 4.6 shows this trend from recent data for average daytime roadway traffic noise exposures between 55 and 80 dBA. This shows an elevated risk beginning at 65 dBA average exposure level.

The first large study on the link between environmental noise exposure and stroke was performed recently in Denmark. With a group of more than 57,000 participants over 5 years, the study revealed an IRR of 1.14 (95% CI: 1.03–1.25) for each increase of 10 dBA in roadway traffic noise between L_{den} 60 and 75 dBA [61]. This correlation could only be made for those above 64.5 years of age and no such correlations were found for rail or

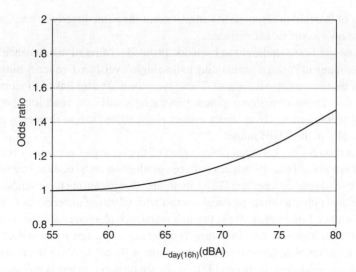

Figure 4.6 Dose–response relationship between average daytime roadway traffic noise and ischemic heart disease (adapted from equation in Babisch 2008 [60])

aircraft noise. Confounding factors in this study included age, gender, smoking, diet (in terms of fruit and vegetable intake), income, physical activity, body mass index, alcohol consumption, coffee consumption, and air pollution (NO_x concentrations).

4.5 Vibroacoustic Disease

Vibroacoustic disease (VAD) is a condition documented almost exclusively by a research group in Portugal, beginning with a group of articles published in 1999 in a single publication [62–71]. This condition was first diagnosed for a group of 140 male aircraft maintenance workers who were exhibiting similar symptoms, most notably lesions throughout the body and thickening of the pericardium and mitral valve of the heart. The pericardium is the double-layered tissue that encases and protects the heart, and the mitral valve facilitates blood flow into the left ventricle. The cause of VAD is said to be long-term exposure to high-level, low-frequency noise. This is defined as exposures of more than 10 years to sound pressure levels exceeding 90 dB in the frequency range below 500 Hz.

If VAD is a valid condition, it is a purely physiologically based noise-induced condition as, unlike the cardiovascular effects mentioned earlier, psychological stress is not involved in its process. There are reportedly three clinical stages of VAD [64]. Stage I VAD is said to result from around 2 years' exposure and is characterized by mood swings, gastrointestinal dysfunction, and respiratory infections. Stage II VAD is said to result from 5–10 years' exposure and adds chest and back pain, skin infections, and dysuria (painful urination). Stage III VAD is said to result from more than 10 years' exposure and adds psychiatric disturbances, hemorrhages, ulcers, colitis, headaches, visual issues, severe pain, neurological disturbances (including epilepsy [66] and balance issues [67]), and thickening of cardiac structures. Lesions in the brain allegedly caused by VAD have been linked with palmomental reflex [68] (a twitching of the chin in response to stimulation of the palm directly below the thumb, in the region known as the thenar eminence).

Limited, isolated case studies have been documented on aircraft maintenance workers at a specific company in Portugal, attempting to link high-level, low-frequency noise exposure to the list of diseases mentioned above. These symptoms are also linked to vibration exposures from direct contact with equipment generating significant vibration levels, [70,71] although vibrational energy from direct contact is not related to low-frequency noise exposures, especially in the 90 dB range.

An explanation for the common presentation of thickening of coronary structures from VAD is that vibrational energy induces collagen production in the body to strengthen structures that are weakened by vibration [72]. Vibrational energy causing these kinds of reactions would be related either to direct physical contact with vibrating materials or to exposure to sound levels much higher than 90 dB through resonance phenomena, as mentioned at the beginning of this chapter. Recent literature from the same research group in Portugal is now attempting to extrapolate low-frequency exposures < 90 dB to VAD for residents living near wind farms and industrial plants [73]. As the sound pressure levels referenced in these recent publications are comparable to those most people experience in typical urban and suburban environments, an explanation is needed to account for why these specific sound sources elicit VAD in some people, while others exposed to similar levels from different sources do not experience these symptoms. It should also be noted that the authors of these studies state that no dose–response relationships have been established for VAD and that their research needs to be supported by further and larger studies. To-date, the scientific and medical communities have not recognized VAD as a legitimate disease for diagnosis [74].

4.6   Low-frequency Noise Concerns

Low-frequency sound is generally considered to have dominant acoustic energy in the range of 20–200 Hz. As discussed in Chapter 3, human hearing sensitivity is reduced in this frequency range compared with that in our range of highest sensitivity (500–8,000 Hz). As is shown in Figure 3.3, the threshold of hearing at 100 Hz is roughly 27 dB and the threshold at 20 Hz is nearly 80 dB. Although there is general agreement that high-levels (over 110 dB) of low-frequency sound can be associated with physiological responses, there is debate about the potential effects associated with moderate levels. The strongest arguments for the effects of low-frequency noise are in the psychological realm and they are addressed in Chapter 5.

Sounds having significant low-frequency components are much more prevalent in contemporary society than they have ever been, due to the expansion of technology throughout the world. Industrial facilities and transportation sources generate low-frequency noise in the environment, while mechanical ventilation systems, personal listening devices, and household tools and appliances generate low-frequency noise in the home and office. In addition to these, recreational sources such as concerts, nightclubs, power tools, and firearms (for hunting and target-shooting) can generate significant levels of low-frequency noise. Sounds in this frequency range are also the most difficult to reduce by normal noise control methods.

Our diminished hearing sensitivities for low-frequency sounds combined with the lack of regulation of sounds in this frequency range can leave us more vulnerable to high levels of low-frequency sound. The key question, however, is what constitutes a physical risk, and this is where there is no consensus for moderate levels. Yet another complicating factor in this mix is the fact that low-frequency noise can generate feelings of vibration in the human body, as mentioned in Chapter 3, with feeling thresholds of 82 dB at 63 Hz and 103 dB at 20 Hz [75]. These thresholds are lower for the head, being 61 dB at 63 Hz and 88 dB at 20 Hz [76], less than 10 dB above the hearing threshold between 20 and 40 Hz [77]. Although the perception of vibration does not imply any potential for negative physiological effects, that sensation can add to stress reactions. This is discussed further in Chapter 5.

The physiological reaction to stress is demonstrated through the activities of the endocrine system in the release of hormones such as cortisol. The fact that moderate levels of low-frequency noise are more annoying to people than mid-frequency noise has been demonstrated through studies monitoring cortisol concentrations in work environments [78]. The condition that the hearing mechanism is always functioning, even during sleep, as a protective measure can result in sleep disturbance from sounds interpreted as threatening. Sleep disturbance, from any source, is generally acknowledged as a cause of compromise in the immune system which can lead to a host of illnesses. The evidence of pure physiological effects related to low-frequency noise is limited. Most effects associated with low-frequency noise are psychological and these are discussed in Chapter 5.

Phantom sounds, those that can be heard by some but not by all and have no obvious physical source associated with them, are often associated with low frequencies. These have been publicized over the past 20 years through reports in the media, and have been labeled as "hums." Some attribute the hearing of such sounds to a form of tinnitus, [79] but there are several other potential explanations for these sensations. Hums are discussed in more detail in Chapter 6.

A physiological effect that has been linked to high-level (>90 dB) low-frequency noise exposure is endolymphatic hydrops [80], a condition in which the chambers of scala media in the inner ear are swollen from excessive endolymphatic fluids (as mentioned in Chapter 3). In an experiment with guinea pigs, a 140 Hz tone delivered in bursts between 88 and 112 dB resulted in clear visual distension of the Reissner's membrane [81]. This has been shown to affect potassium concentrations in these fluids, which can affect the electrical signals sent over the auditory vestibular nerve to the brain. As the vestibule and semicircular canals of the inner ear are also filled with endolymphatic fluids, this can affect balance as well as hearing. Recovery from this condition is generally quick after the sound exposure ceases but, for some illnesses, such as Ménière's disease, endolymphatic hydrops has no relation to sound exposure, resulting in hearing loss and vertigo for extended periods.

Although the prevailing sentiment is that sound below the threshold of hearing produces no physiological effects in people, recent research is showing the potential for low-frequency sound below the threshold of hearing to have some effect [82]. The mechanism through which this is stated to occur is greater stimulation of the outer hair cells by low-frequency sound. As mentioned in Chapter 3, most of the information conveyed to the brain from the inner ear is carried through the inner hair cells, but the inner hair cells are not in direct contact with the tectorial membrane, thus reducing the conveyance of low-frequency energy information. The outer hair cells are in direct contact with the tectorial membrane, so more low-frequency information can be conveyed through them in ways not previously considered. This information appears to be conveyed through the hearing system (as measured by electrical signals between the inner ear and the brain) but somehow discarded. The hypothesis behind this phenomenon is that these sounds are meant to be inhibited from our consciousness to avoid the potential distractions associated with hearing internal low-frequency sounds such as those associated with breathing, heartbeat, and body movements [83].

This information is not completely discarded, though, as has been shown by monitoring otoacoustic emissions from the cochlea following high-level, low-frequency sound exposures. A recent German study exposing subjects to a 30 Hz tone at 120 dB sound pressure level revealed more cochlear activity than expected as demonstrated through otoacoustic emissions from outer hair cells [84]. Although the study authors state that the exposure level is moderate and not intrusive when they translate the 120 dB at 30 Hz to an A-weighted level of 80 dBA (which would be considered to be moderate if dominated by mid-frequency energy), a 120 dB sound pressure level at 30 Hz is well above the threshold of hearing and considered to be very unpleasant by most people. This is an example of potentially misleading information, as the premise is that common moderate levels of low-frequency sound can evoke significant responses in the hearing mechanism, while the level of sound tested is actually high and not common. Common, moderate levels of low-frequency sound would be 40–50 dB lower than that used in this study.

In contrast to the equal loudness curves shown in Figure 2.3, Figure 4.7 shows a family of equal-unpleasantness curves for low-frequency sound based on an equation published in a recent study [85]. This was based on a laboratory experiment in which 39 adult subjects were exposed to pure tones and asked to judge the acceptability of those tones in different settings, such as a living room, bedroom, office, and factory. For comparison, the "very unpleasant" curve from Figure 4.7 is slightly higher than the 60 phon curve from Figure 2.3 and the "unpleasant" curve is roughly equivalent to the 40 phon curve in the low-frequency range. As this chart shows, the 120 dB signal at 30 Hz from the German study would be above the "very unpleasant" curve.

4.7   Infrasound Concerns

Infrasound is generally considered to have dominant acoustic energy below 16–20 Hz. Although some references state that infrasound is not audible, the most accurate description of infrasound is that it is audible at levels more than 80 dB higher than that for sounds with predominant energy in the mid-frequency range (500–4,000 Hz). In fact, stimulation in the 1 Hz range has been shown to generate significant endolymph movement and cochlear hair cell activity [86].

It has been suggested that, due to the plethora of misunderstandings and negative connotations associated with infrasound, what is mostly characterized as infrasound should be included in the category of low-frequency sound [87]. Some references published in the 1960s made extreme, unsubstantiated claims about effects that moderate levels of infrasound can

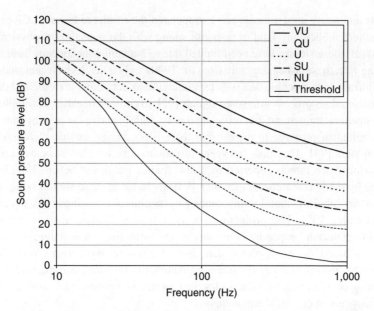

Figure 4.7 Equal-unpleasantness curves for low-frequency sound. NU, not unpleasant; SU, somewhat unpleasant; U, unpleasant; QU, quite unpleasant; VU, very unpleasant; threshold, threshold of hearing. (Adapted from equation in Inukai et al. [85])

have on people, and to this day these publications have resulted in significant public concern. Tales of debilitating effects related to infrasound have led to folklore about acoustic weapons that have yet to be proven. Vlademir Gavreau wrote about a colleague who was rendered "a lifelong invalid" from his exposure to infrasound, although the frequency of the source was referenced as 2,600 Hz and the sound level was referenced in terms of the power of the source (1 kW) rather than the sound pressure level at the position of the listener [88].

Infrasound has many natural causes, such as earthquakes, volcanoes, ocean wave motion, avalanches, meteors, lightning, and high winds [89]. Monitoring of infrasound can be instrumental in early warning systems for natural disasters resulting from these kinds of activities, especially since the energy associated with sound waves of such low frequency is barely attenuated with distance (unlike energy associated with higher frequencies). This explains the crack one hears near a lightning strike in comparison with the rumble heard at a distance. Although there have been isolated attempts to link natural infrasound to physiological effects, there is no consensus on the validity of these claims. A few studies in the 1960s attempted to link pre-Föhn weather conditions with automobile accidents, mortality, and birth rates in Austria and Switzerland [90,91]. Föhn weather is caused by low-pressure systems moving over mountain ranges. In addition to sharp changes in temperature, barometric pressure, and humidity associated with Föhn weather conditions near the Alps, high winds are often associated with these phenomena, which can generate significant levels of infrasound.

Several studies have been performed over the past 50 years to define the limits of physiological effects associated with infrasound exposures. Each study used a limited number of subjects and many used animals and extrapolated the results to humans. Given the high levels required to elicit any kind of response, along with the potential for physical damage above 170 dB, these studies have been limited due to humanitarian concerns, and the literature dealing with these issues is mostly from the 1980s and earlier. Most recent studies on the effects of infrasound on people deal with psychological effects, but one from 2007 discovered that the cochlea does process infrasonic signals [92], in contrast to earlier assumptions.

There is general agreement that infrasound only slightly above the hearing threshold can induce a feeling of pressure in the ear, but there is no consensus that such levels induce other symptoms [93]. The links between infrasound exposure levels below 130 dB and human performance effects reported in a few studies published before 1980 have been questioned by many researchers since then [94]. Included in these studies was one linking interior automobile infrasound to involuntary nystagmus and feelings typically associated with intoxication [95]. The problem with most studies linking negative physiological effects with infrasound exposure is the mix of unsubstantiated anecdotal information and assumptions based on extrapolated data from legitimate experiments involving much higher level exposures than those referenced. Another issue with these studies is the lack of consideration of other potential causes for the reported symptoms, such as electromagnetic hypersensitivities or chemical exposures.

One aspect of infrasound that has gained some attention is the potential for it to be linked with paranormal activities or haunted areas. A few, fairly limited, studies reporting on strange feelings experienced in some buildings linked them with low-level (< 40 dB) 19 Hz signals, while the sightings of ghosts were linked to eyeball resonances [96,97]. These studies have been refuted by more recent publications, which note the greater potential for the referenced feelings to result from natural and man-made electromagnetic fields than from infrasound [98]. Also, the reference to eyeball resonance causing the sighting of apparitions off to the side has been refuted due to the combination of the low levels of infrasound reported and the notion that eyeball resonance would affect the entire visual field rather than just one side. A recent controlled study was performed in which 19 and 22 Hz signals (to replicate the frequencies associated with earlier studies) at 75 dB were introduced separately and in combination with an electromagnetic signal used by other investigations. This yielded no conclusive evidence of a link between infrasound and these experiences, with some evidence that electromagnetic fields might play a role [99].

References

[1] Tempest, W. *Infrasound and Low Frequency Vibration*. London: Academic Press, 1976.
[2] Hakansson, B., Brandt, A. and Carlsson, P. (1994). "Resonance frequencies of the human skull in vivo." *Journal of the Acoustical Society of America*, 95(3): 1474–1481.
[3] Mohr, G.C., et al. (1965). "Effects of low frequency and infrasonic noise on man." *Aerospace Medicine*, 36(9): 817–824.
[4] Schmiedt, R.A. (2010). "The physiology of cochlear presbycusis." In: *The Aging Auditory System*. Springer New York, pp. 9–38.
[5] Dubno, J.R., et al. (2008). "Longitudinal changes in speech recognition in older persons." *Journal of the Acoustical Society of America*, 123(1): 462–475.

[6] Lee, F.S., et al. (2005). "Longitudinal study of pure-tone thresholds in older persons." *Ear and Hearing*, 26: 1–11.

[7] International Organization for Standardization. *ISO 7029:2000. Acoustics – Statistical distribution of Hearing Thresholds as a Function of Age*. Geneva: ISO, 2000.

[8] Rosen, S., et al. (1962). "Presbycusis study of a relatively noise-free population in the Sudan." *The Annals of Otology, Rhinology & Laryngology*, 71:727–743.

[9] Kryter, K.D. (1983). "Presbycusis, sociocusis and nosocusis." *Journal of the Acoustical Society of America*, 73(6): 1897–1917.

[10] Kurakata, K. (2008). "Effect of ageing on hearing thresholds in the low frequency region." *Journal of Low Frequency Noise, Vibration and Active Control*, 27(3): 175–184.

[11] Fosbroke, C. and Fosbroke, J. (1831). "Practical observations on the pathology and treatment of deafness." *Lancet*, VI: 645–648.

[12] Barr, T. (1886). "Enquiry into the effects of loud sounds upon the hearing of boilermakers and others who may work amid noisy surroundings." *Proceedings of the Philosophical Society of Glasgow*, 18: 223–239.

[13] Holt, E.E. 1882. "Boiler-maker's deafness and hearing in noise." *Transactions of the American Otological Society*, 3:34–44.

[14] Berglund, B., Lindvall, T., and Schwela, D.H. *Guidelines for Community Noise*. Geneva: World Health Organization, 1999.

[15] Saunders, J.C., Dear, S.P. and Schneider, M.E. (1985). "The anatomical consequences of acoustic injury: A review and tutorial." *Journal of the Acoustical Society of America*, 78(3): 833–860.

[16] Henderson, D. and Hamernik, R.P. (1995). "Biologic basics of noise induced hearing loss." *Occupational Medicine: State of the Art Reviews*, 10(3): 513–534.

[17] Talaska, A.E. and Schacht, J. (2007). "Mechanisms of noise damage to the cochlea." *Audiological Medicine*, 5: 3–9.

[18] Taylor, W., et al. (1965). "Study of noise and hearing in jute weaving." *Journal of the Acoustical Society of America*, 38(1): 113–120.

[19] International Organization for Standardization. *ISO 1999:2013. Acoustics – Estimation of noise-induced hearing loss*. Geneva: ISO, 2013.

[20] Henderson, D., Subramaniam, M., and Boettcher, F.A. (1993). "Individual susceptibility to noise-induced hearing loss: An old topic revisited." *Ear and Hearing*, 14(3): 152–168.

[21] Wong, A.C.Y., Froud, K.E. and Hsieh, Y.S-Y. (2013). "Noise-induced hearing loss in the 21st century: A research and translational update." *World Journal of Otorhinolaryngology*, 3(3): 58–70.

[22] Prasher, D. and Luxon, L. *Advances in Noise Research Volume 1: Biological Effects of Noise*. London: Whurr Publishers Ltd, 1998.

[23] Barrenas, M.L. and Lindgren, F. (1991). "The influence of eye color on susceptibility to TTS in humans." *British Journal of Audiology*, 25: 303–307.

[24] Ward, W.D. (1995). "Endogenous factors related to susceptibility to damage from noise." *Occupational Medicine*, 10(3): 561–575.

[25] Fechter, L.D, Young, J.S. and Carlisle, L. (1988). "Potentiation of noise induced threshold shifts and hair cell loss by carbon monoxide." *Hearing Research*, 34(1): 39–47.

[26] Henderson, D., et al. (1994). "The role of middle ear muscles in the development of resistance to noise induced hearing loss." *Hearing Research*, 74:22–28.

[27] Canlon, B., Borg, E., and Flock, A. (1988). "Protection against noise trauma by pre-exposure to a low-level acoustic stimulus." *Hearing Research*, 34: 197–200.

[28] Campo, P., et al. (1991). "The effect of 'conditioning' exposures on hearing loss from traumatic exposure." *Hearing Research*, 55: 195–200.

[29] Ryan, A.F., et al. (1994). "Protection from noise induced hearing loss by prior exposure to a nontraumatic stimulus: role of the middle ear muscles." *Hearing Research*, 72:23–28.

[30] Subramaniam, M. Henderson, D., and Spongr, V. (1993). "Effect of low-frequency "conditioning" on hearing loss from high-frequency exposure." *Journal of the Acoustical Society of America*, 93(2): 952–956.

[31] Clark, W.W., Bohne, B.A. and Boettcher, F.A. (1987). "Effect of periodic rest on hearing loss and cochlear damage following exposure to noise." *Journal of the Acoustical Society of America*, 82(4): 1253–1264.

[32] Niskar, A., et al. (2001). "Estimated prevalence of noise-induced hearing threshold shifts among children 6 to 19 years of age: the third national health and nutrition examination survey, 1988–1994, United States." *Pediatrics*, 108(1): 40–43.

[33] Henderson, E., Testa, M.A. and Hartnick, C. (2011). "Prevalence of noise-induced hearing-threshold shifts and hearing loss among US youths." *Pediatrics*, 127(1): e39-e46.

[34] Westcott, M. (2006). "Acoustic shock injury (ASI)." *Acta Otolaryngologica*, 126: 54–58.

[35] Joachims, Z., et al. (1983). "Dependence of noise-induced hearing loss upon perilymph magnesium concentration." *Journal of the Acoustical Society of America*, 74(1): 104–108.

[36] Kopke, R.D., et al. (2007). "NAC for noise: from bench top to the clinic." *Hearing Research*, 226: 114–125.

[37] Lynch, E.D. and Kil, J. (2005). "Compounds for the prevention and treatment of noise induced hearing loss." *Drug Discovery Today*, 10(19): 1291–1298.

[38] Brignull, H.R., Raible, d.W., and Stone, J.S. (2009). "Feathers and fins: Non-mammalian models for hair cell regeneration." *Brain Research*, 1277: 12–23.

[39] Phaneuf, R. and Hetu, R. (1990). "An epidemiological perspective of the causes of hearing loss among industrial workers." *The Journal of Otolaryngology*, 19(1): 31–40.

[40] Morata, T.C., et al. (1993). "Effects of occupational exposure to organic solvents and noise on hearing." *Scandinavian Journal of Work Environment and Health*, 19: 245–254.

[41] Fuente, A. and McPherson, B. (2006). "Organic solvents and hearing loss: The challenge for audiology." *International Journal of Audiology*, 45(7): 367–381.

[42] Morata, T.C. (2007). "Promoting hearing health and the combined risk of noise-induced hearing loss and ototoxicity." *Audiological Medicine*, 5(1): 33–40.

[43] Rybak, L.P. (1992). "Hearing: the effects of chemicals." *Otolaryngology Head Neck Surgery*, 106(6): 677–686.

[44] Sass-Kortsak, A.M., Corey, P.N. and Robertson, J.M. (1995). "An investigation of the association between exposure to styrene and hearing loss." *Annals of Epidemiology*, 5(1): 15–24.

[45] Jacobsen, P., et al. (1993). "Mixed solvent exposure and hearing impairment: an epidemiological study of 3284 men. The Copenhagen male study." *Occupational Medicine (Oxf)*, 43: 180–184.

[46] Tomei, F., et al. (1995). "Epidemiological and clinical study of subjects occupationally exposed to noise." *International Journal of Angiology*, 4: 117–121.

[47] Ising, H. and Braun, C. (2000). "Acute and chronic endocrine effects of noise: Review of the research conducted at the Institute for Water, Soil and Air Hygiene." *Noise & Health*, 7:7–24.

[48] Melamed, S., Fried, Y., and Froom, P. (2001). "The interactive effect of chronic exposure to noise and job complexity on changes in blood pressure and job satisfaction: a longitudinal study of industrial employees." *Journal of Occupational Health Psychology*, 6(3):182–195.

[49] Jarup, L., et al. (2008). "Hypertension and exposure to noise near airports: the HYENA study." *Environmental Health Perspectives*, 116(3):329–333.

[50] Babisch, W., et al. (2012). "Exposure modifiers of the relationships of transportation noise with high blood pressure and noise annoyance." *Journal of the Acoustical Society of America*, 132(6): 3788–3808.

[51] van Kempen, E. and Babisch, W. (2012). "The quantitative relationship between road traffic noise and hypertension: a meta-analysis." *Journal of Hypertension*, 30(6): 1075–1086.

[52] Dratva, J., et al. (2012). "Transportation noise and blood pressure in a population-based sample of adults." *Environmental Health Perspectives*, 120(1): 50–55.

[53] Belojevic, G., et al. (2008). "Urban road-traffic noise and blood pressure in school children." *Proceedings of ICBEN 2008*, Foxwoods, CT.

[54] Babisch, W. and van Kamp, I. (2009). "Exposure-response relationship of the association between aircraft noise and the risk of hypertension." *Noise & Health*, 11(44): 161.

[55] Babisch, W. (2000). "Traffic noise and cardiovascular disease: epidemiological review and synthesis." *Noise & Health*, 2(8): 9–32.

[56] Babisch, W. (2006). "Transportation noise and cardiovascular risk: Updated review and synthesis of epidemiological studies indicate that the evidence has increased." *Noise & Health*, 8:30.

[57] Babisch, W., et al. (2005). "Traffic noise and risk of myocardial infarction." *Epidemiology*, 16(1): 33–40.

[58] Sørensen, M., et al. (2012). "Road traffic noise and incident myocardial infarction: a prospective cohort study." *PloS One*, 7(6): e39283.

[59] Gan, W.Q., et al. (2012). "Association of long-term exposure to community noise and traffic-related air pollution with coronary heart disease mortality." *American Journal of Epidemiology*, 175(9): 898–906.

[60] Babisch, W. (2008). "Road traffic noise and cardiovascular risk." *Noise & Health*, 10(38): 27–37.

[61] Sørensen, M., et al. (2011). "Road traffic noise and stroke – A prospective cohort study." *European Heart Journal*, 32(6): 734–744.

[62] Castelo Branco, N.A.A and Rodriguez, E. (1999). "The vibroacoustic disease – An emerging pathology," *Aviation, Space, and Environmental Medicine*, 70(3), Section II: A1-A6.

[63] Alves-Pereira, M. (1999). "Noise-induced extra-aural pathology: A review and commentary." *Aviation, Space, and Environmental Medicine*, 70(3), Section II: A7–A21.

[64] Castelo Branco, N.A.A. (1999). "The clinical stages of vibroacoustic disease." *Aviation, Space, and Environmental Medicine*, 70(3), Section II: A32–39.

[65] Marciniak, W., et al. (1999). "Echocardiographic evaluation in 485 aeronautical workers exposed to different noise environments." *Aviation, Space, and Environmental Medicine*, 70(3), Section II: A46–A53.

[66] Martinho Pimenta, A.J.F. and Castelo Branco, N.A.A. (1999). "Neurological aspects of vibroacoustic disease." *Aviation, Space, and Environmental Medicine*, 70(3), Section II: A91–A95.

[67] Martinho Pimenta, A.J.F., Castelo Branco, M.S.N. and Castelo Branco, N.A.A. (1999). "Balance disturbances in individuals with vibroacoustic disease." *Aviation, Space, and Environmental Medicine*, 70(3), Section II: A96–A99.

[68] Martinho Pimenta, A.J.F., Castelo Branco, M.S.N. and Castelo Branco, N.A.A. (1999). "The palmo-mental reflex in vibroacoustic disease." *Aviation, Space, and Environmental Medicine*, 70(3), Section II: A100-A106.

[69] Martinho Pimenta, A.J.F. and Castelo Branco, N.A.A. (1999). "Facial dyskinesia induced by auditory stimulation: A report of four cases." *Aviation, Space, and Environmental Medicine*, 70(3), Section II: A119-A121.

[70] Martinho Pimenta, A.J.F. and Castelo Branco, N.A.A. (1999). "Epilepsy in the vibroacoustic disease: A case report." *Aviation, Space, and Environmental Medicine*, 70(3), Section II: A122-A127.

[71] Castelo Branco, N.A.A., et. al. (1999). "Vibroacoustic disease: Some forensic aspects." *Aviation, Space, and Environmental Medicine*, 70(3), Section II: A145-A151.

[72] Alves-Pereira, M. and Castelo Branco, N.A.A. (2007). "Vibroacoustic disease: Biological effects of infrasound and low-frequency noise explained by mechanotransduction cellular signaling." *Progress in Biophysics and Molecular Biology*, 93: 256–279.

[73] Alves-Pereira, M. and Castelo Branco, N.A.A. (2007). "Public health and noise exposure: the importance of low frequency noise." *Proceedings of Internoise 2007*, Istanbul, Turkey.

[74] Von Gierke, H.E. and Mohler, S.R. (2002). "Vibroacoustic disease." *Aviation, Space, and Environmental Medicine*, 73(8): 828–830.

[75] Yamada, S. et al. (1983). "Body sensations of low frequency noise of ordinary persons and profoundly deaf persons." *Journal of Low Frequency Noise and Vibration*, 2(3): 32–36.

[76] Takahashi, Y. (2013). "Vibratory sensation induced by low-frequency noise: The threshold for "vibration perceived in the head" in normal-hearing subjects." *Journal of Low Frequency Noise, Vibration and Active Control*, 32(1+2): 1–10.

[77] Takahashi, Y. (2009). "Vibratory sensation induced by low-frequency noise: A pilot study on the threshold level." *Journal of Low Frequency Noise, Vibration and Active Control*, 28(4): 245–253.

[78] Persson-Waye, K., et al. (2002). "Low frequency noise enhances cortisol among noise sensitive subjects during work performance." *Life Sciences*, 70: 745–758.

[79] Van den Berg, F. (2009). "Low frequency noise and phantom sounds." *Journal of Low Frequency Noise, Vibration and Active Control*, 28(2): 105–116.

[80] Salt, A.N. (2004). "Acute endolymphatic hydrops generated by exposure of the ear to nontraumatic low frequency tone." *Journal of the Association for Research in Otolaryngology*. 5: 203–214.

[81] Flock, A. and Flock, B. (2000). "Hydrops in the cochlea can be induced by sound as well as by static pressure." *Hearing Research*, 150: 175–188.

[82] Salt, A.N. and Lichtenhan, J.T. (2012). "Perception-based protection from low-frequency sounds may not be enough." *Proceedings of Internoise 2012*, New York.

[83] Salt, A.N., et al. (2013). "Large endolymphatic potentials from low-frequency and infrasonic tones in the guinea pig." *Journal of the Acoustical Society of America*, 133(3): 1561–1571.

[84] Kugler, K., et al. (2014). "Low-frequency sound affects active micromechanics in the human inner ear." *Royal Society Open Science*, 1: 140–166.

[85] Inukai, Y., Nakamura, N. and Taya, H. (2000). "Unpleasantness and acceptable limits of low frequency sound." *Journal of Low Frequency Noise, Vibration and Active Control*, 19(3): 135–140.

[86] Salt, A.N. and DeMott, J.E. (1999). "Longitudinal endolymph movements and endochochlear potential changes induces by stimulation at infrasonic frequencies." *Journal of the Acoustical Society of America*, 106(2): 847–856.

[87] Leventhall, G. (2007). "What is infrasound?" *Progress in Biophysics and Molecular Biology*, 93: 130–137.

[88] Gavreau, V. (1968). "Infrasound." *Science Journal*, 4(1): 33–37.

[89] Bedard, A.J. and Georges, T.M. (2000). "Atmospheric infrasound." *Physics Today*, 53(3): 32–37.

[90] Moos, W.S. (1963). "Effects of 'Fohn' weather on the human population in the Principality of Lichtenstein." *Aerospace Medicine*, 34: 736–739.

[91] Moos, W.S. (1964). "Effects of 'Fohn' weather on accident rates in the City of Zurich (Switzerland)." *Aerospace Medicine*, 35: 643–645.

[92] Hensel, J., et al. (2007). "Impact of infrasound on the human cochlea." *Hearing Research*, 233: 67–76.

[93] Møller, H. (1984). "Physiological and psychological effects of infrasound on humans." *Journal of Low Frequency Noise and Vibration*, 3(1).

[94] Harris, C.S., Sommer, C., and Johnson, D.L. (1976). "Review of the effects of infrasound on man." *Aviation, Space, and Environmental Medicine*, 47(4): 430–434.

[95] Evans, M. J. and Tempest, W. (1972). "Some effects of infrasonic noise in transportation." *Journal of Sound and Vibration*, 22(1): 19–24.

[96] Tandy, V. and Lawrence, T.R. (1998). "The ghost in the machine." *Journal of the Society for Psychical Research*, 62(851): 360–364.

[97] Tandy, V. (2000). "Something in the cellar." *Journal of the Society for Psychical Research*, 64(860): 129–140.

[98] Braithwaite, J. and Townsend, M. (2006). "Good vibrations: the case for a specific effect of infrasound in instances of anomalous experience has yet to be empirically demonstrated." *Journal of the Society of Psychical Research*, 70.4(885): 211–224.

[99] French, C.C, et al. (2009). "The 'Haunt' project: An attempt to build a 'haunted' room by manipulating complex electromagnetic fields and infrasound." *Cortex*, 45: 619–629.

5

Psychological Effects of Sound Exposure

5.1 Introduction

Most of the effects of sound on people, both positive and negative, are psychologically based. Most physiological effects of sound exposure are also psychologically based, as they result from subjective interpretations of the sounds, which, in turn, result in stress or pleasure. It is stress resulting from the negative interpretation of a sound that causes the negative physiological effect, rather than the mere exposure to the sound. The only exception to this is related to the extremely high levels that can cause physical damage to the organs of the hearing mechanism and other body tissues, independent of their psychological interpretation, as described in Chapter 4. As most sound exposures are well below physically damaging levels, psychological effects from sound are much more common than physiological effects and, therefore, much more research has been performed in the psychological arena. The principal negative interpretations of sound exposure are annoyance and stress, and the by-products of these that have been studied the most are sleep disturbance, learning disabilities, and emotional states. This chapter summarizes the most recent research in these areas, with an emphasis on negative effects. The positive effects of sound exposure are discussed in more detail in Chapter 7.

5.2 Annoyance

Although annoyance is subjective and can only be rated statistically, there are almost universally annoying types of sounds, such as the screech associated with metal-on-metal friction or fingernails scraping against a blackboard. An experiment performed to measure the frequency content of such sounds revealed a mix of energy peaks at 1,400 Hz and harmonics (integer multiples) of that frequency out to 7,000 Hz [1]. By varying sound levels and filtering out different parts of the spectra while having subjects rate the unpleasantness of the sound, it was revealed that the sound is equally annoying regardless of the level, but less annoying when the components below 4,000 Hz were removed. Therefore, the 1,400 and 2,800 Hz components caused the most annoyance of the acoustic signature.

Common annoyance increases with increasing sound pressure level (SPL), but annoyance is also a function of subjective context. The most prevalent attitudinal causes for noise annoyance are fear of danger from the source, perceived lack of control of the source, interference with activities, and general noise sensitivity. A noise source viewed as being important tends to decrease the associated annoyance [2]. Another factor that contributes to decreased annoyance is the perceived personal benefit associated with the source [3]. The term "annoyance" is vague enough that there is wide variation even among acoustical experts regarding its definition [4]. The most widely used method for rating annoyance is through arbitrary scales based on attitudinal surveys.

For a specific sound level, mid-frequencies (500– 2,000 Hz) tend to generate significant annoyance, for which reason the dBA scale is used most often to rate human annoyance. However, low-frequency noise is also rated by many as being annoying at lower levels than their counterparts in the mid-frequency range and experiments have shown that the rate of annoyance increases more rapidly with increasing SPL for frequencies below 100 Hz than for frequencies above 100 Hz [5–7]. It has been shown that, for comparable A-weighted SPLs in the 62–84 dBA range, low-frequency noise (between 10 and 250 Hz) is more annoying (by a factor of 8–12%) than non-tonal noise in the mid-frequency range [8]. Considering the reduced sensitivity of the hearing mechanism to low-frequency sound and consequently the dBA scale, the perceived level of low-frequency sound qualified as annoying is much lower than the corresponding mid-frequency level.

There is therefore an ongoing debate about the appropriateness of the use of the dBA scale to rate annoyance. In this regard, several studies have shown that the dBA rating system underestimates the annoyance of low-frequency sounds by 3–8 dB [9,10]. Although some references support the use of dBC or unweighted sound level rather than dBA for rating annoyance, these studies result in annoyance overestimates by 10 dB or more when dBC or unweighted levels were used. Even with these limitations, the dBA scale is used most widely for rating human annoyance.

As annoyance is subjective and cannot be measured with an instrument, it is difficult to rate using consistent descriptors. Schultz, in 1978, published an attempt to quantify noise annoyance in terms of the percentage of people who are "highly annoyed" (%HA) by different transportation sources [11]. He developed curves based on 11 social surveys performed in the 1960s and 1970s in the US and Europe, plotting summary curves of %HA against the day–night equivalent level (L_{dn}), defining %HA as the responses in the top 27–29% annoyance ratings based on 7- and 11-step annoyance scales used in the surveys. These curves have been updated since Schultz's publication, based on more recent surveys [12,13]. Mathematical exposure–response (or dose–response) relationships were developed to correlate annoyance on a 0–100 scale. Three levels of annoyance were defined in this compilation:

- Highly annoyed (%HA) – defined as ratings of 72 or above
- Annoyed (%A) – defined as ratings of 50 or above
- A little annoyed (%LA) – defined as ratings of 28 or above.

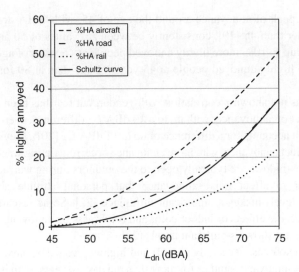

Figure 5.1 Dose–response annoyance curves for transportation sources. %HA, percentage of people "highly annoyed", in this case by transportation noise. (Adapted from equations in Miedema and Oudshoorn [14] and Schultz [11].)

The most recent set of curves is shown in Figure 5.1, derived from 20 aircraft studies, 18 roadway traffic studies, and nine railway studies performed in Australia, North America, and Europe from the 1960s to the 1990s [14]. The single average data compilation developed by Schultz in his seminal 1978 paper is often referred to as the Schultz curve, which is similar to the "road" curve in Figure 5.1 (as is shown in the figure), so these curves have not changed significantly since they were first introduced. These studies show that aircraft noise is considered to be the most annoying of transportation sources and rail noise is the least annoying for the same average sound level exposure in the US and Europe. Some more recent studies in Korea and Japan have shown that rail noise is considered to be more annoying than road noise [15], but this appears to be caused by the closer proximity of rail operations to residences in Korea and Japan (along with different attitudes towards the sources) compared with conditions in the US and Europe [16].

The studies described addressed exposures to each category of sound source separately; however, several sources are often combined in some environments. A Swedish study addressed this issue by surveying close to 2,000 adults (aged 18–75 years) exposed to transportation noise with a 24-hour equivalent sound level [$L_{eq(24)}$] of between 45 and 72 dBA. Those exposed to combined rail and roadway noise voiced significantly more annoyance than those exposed to only rail or roadway noise at the same sound levels for exposures exceeding 58 dBA [17].

Other annoyance study compilations have been performed to define trends with demographics. A study on close to 3,000 primary school children and their parents in Europe showed that, although children are less annoyed than their parents about daytime (7:00 am to 11:00 pm) aircraft and traffic L_{eq} exposures above 55 dBA, their reactions are broadly comparable [18]. A recent compilation of international transportation noise studies totaling more than 62,000 people between the ages of 15 and 102 revealed an inverted U-shape trend showing a peak of annoyance around the age of 45, with %HA dropping for those younger and older than 45 [19], consistently between L_{dn} values of 50 and 70 dBA. This analysis yielded up to 10% more highly annoyed people at 45 years of age than at 20, and up to 20% more highly annoyed people at 45 years of age than at 80 for the same noise exposure level.

Annoyance has not shown a correlation with gender, but fear has been shown to have a significant impact on annoyance, with up to a 19 dBA L_{dn} difference, even more than noise sensitivity (which accounts for a difference of up to 11 dBA L_{dn} [20]). Meteorology has also been shown to affect annoyance, with more annoyance occurring during warmer months [21], perhaps because windows are typically open to the outdoors during warmer times. Culture does not appear to affect noise annoyance, but personal attitude about the source (independent of level) makes a significant difference [22]. Some researchers have even experimented with the effects of indoor color and brightness on annoyance from noise, but their results have not shown significant trends [23].

Transportation sources, especially aircraft and highway vehicles, have been the subject of most published annoyance studies for more than the past 50 years, with isolated industrial source studies and a recent introduction of several wind farm noise studies over the past 10 years. Figure 5.2 shows the results of some of those studies, with dose–response annoyance curves for wind farm noise. These relationships were derived from three studies from Sweden and the Netherlands in which 1,820 surveys were evaluated using 72 on a 0–100 annoyance scale to define %HA [24]. Although the database for Figure 5.2 is much smaller than that used for Figure 5.1, it is worth noting that wind farm noise is rated as more

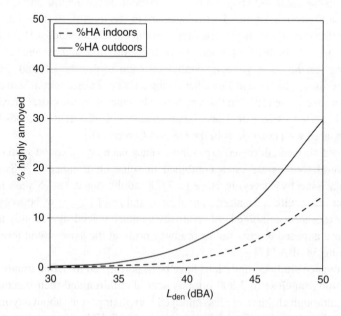

Figure 5.2 Dose–response annoyance curves for wind farms. %HA, percentage of people "highly annoyed". (From Janssen et al. [24] with permission of the Acoustical Society of America.)

annoying than transportation noise, as the onset of annoyance corresponds to sound levels 10 dBA lower for wind farms than for transportation sources. However, noise from wind farms is a more complicated matter than noise from transportation sources, with, for example, lower annoyance ratings associated with those who are benefiting economically from the wind farm and higher annoyance ratings associated with those who can see a wind turbine from inside their home. The topic of wind farm noise is discussed further in Chapter 6.

5.3 Stress

It has long been known anecdotally that noise is associated with psychological stress, but recent research has shown this link through monitoring of key stress hormones secreted by the adrenal gland of the endocrine system. Hormones most associated with noise are cortisol, noradrenaline, and adrenaline [25]. Increased noradrenaline and adrenaline are associated with the fight-or-flight reaction and increased cortisol is associated with a defeat or a learned helplessness reaction. Specific noise levels that trigger stress reactions are difficult to define, as the context and interpretation of the sound can have much more to do with the reaction than the level itself. Another key variable is an individual's sensitivity to different sources and types of noise. As noradrenaline and adrenaline releases are associated with a feeling of danger, a low level of sound associated with a threatening source can produce greater stress reactions than a high level associated with a non-threatening or habituated source. This would occur during both waking and sleeping conditions as the hearing mechanism is always active [26].

Cortisol secretions vary over a 24-hour period due to natural circadian rhythms. The levels are typically high in the morning after waking up and they decrease as the day progresses. Several studies have been performed linking cortisol increase with noise exposures independent of typical variations. High levels of occupational noise exposures (above daily L_{eq} of 85 dBA) have been shown to cause increases in cortisol levels as well as fatigue and irritability. This was clearly demonstrated by testing the same workers in the same environments with and without effective hearing protection (providing at least 30 dBA of attenuation) and noting a decrease in cortisol to normal levels expected over the course of the day [27].

Although there is little doubt about the stress-inducing potential of high-level noise, low and moderate noise levels cause stress based on subjective reactions to the sounds. Research study results are mixed when they are based on personal ratings of the stress caused by different noise sources. A recent Swedish study, based on laboratory exposure to 78 dBA of tractor noise while subjects performed tasks, showed no significant cortisol effects from noise exposure, even though subjective ratings resulted in higher stress ratings [28]. A study in small towns of Austria showed elevated cortisol levels for fourth-grade children exposed to L_{dn} community noise levels > 60 dBA as compared with children of the same age exposed to L_{dn} community noise levels < 50 dBA [29].

Low-frequency noise has also been studied in relation to stress responses. A study in Germany of children aged 7–10 years showed elevated cortisol levels during the first half of the night as a result of trucks passing every 4 minutes within 3–8 m of homes [30]. The first half of the night is when cortisol levels are typically lower due to circadian rhythms, so these elevated levels can disrupt restful sleep. Bear in mind that the children in this study had to be woken up 1 hour after going to sleep to provide urine samples, something that, by itself, might raise cortisol levels.

A laboratory study specifically targeted the effects of low-frequency noise on subjects exposed to 40 dBA of steady noise with and without dominant (>10 dB higher than the reference signal) low-frequency components in the 30–60 Hz range [31]. These exposures were on separate days during the afternoon when cortisol levels are typically decreased. Some of the subjects were also designated as highly sensitive to noise from questionnaires. This yielded elevated cortisol levels only for highly sensitive subjects exposed to the signals with elevated low-frequency noise components [31].

Subjective, self-reported stress among school children has been linked to aircraft noise by studies in the UK [32]. These studies showed no habituation with time to the noise when comparing groups of children aged 8–11 exposed to 16-hour L_{eq} values > 66 dBA and < 57 dBA.

One other aspect of interest with stress reactions is the perceived control over the noise. A recent German study among 30- to 45-year-old women showed that elevated adrenaline and noradrenaline concentrations were evident only when attempts to control the noise (by closing windows to reduce road traffic noise) did not reduce the perceived disturbance [33].

5.4   Sleep Disturbance

As sleep can have a profound effect on health, sleep disturbance by noise has been an active research topic over the past 40 years. Most of this research is related to transportation noise, predominantly aircraft and roadway traffic. Dose–response relationships have been developed for different aspects of sleep disturbance by noise, but it is important to identify the real source of the sleep disturbance before linking it exclusively to a noise source. Sleep can be disturbed by internal as well as external agents. Internal agents include the effects of medical conditions, medications, and anxiety about things that are unrelated to noise, such as safety, work environment, health, finances, and personal relationships. Noise would be included in the external agent category, and personal sensitivity to the specific noise source makes a significant difference in responses to the noise.

There are many potential confounding factors in the attempts to relate sleep disturbance to noise, including those already mentioned along with personal sleep patterns and sensitivities. Another key aspect to linking sleep disturbance to noise is the method by which sleep disturbance is determined. Sleep disturbance studies use several methods in their evaluations, most common of which are actimetry, electroencephalography (EEG) monitoring, behavioral response monitoring, and questionnaires. Actimeters are devices typically worn on wristbands that monitor body movements, EEG monitors brain activity, behavioral responses are typically monitored by the subject pressing a button upon awakening, and questionnaires provide subjective assessments of the degree of disturbance. Each of these methods yields different results in attempting to assign specific noise level limits to sleep disturbance reactions. Beyond that, one must determine what constitutes the state of awakening. These include definitions for the states of arousals, awakenings, and sleep stage changes. Of these, awakenings are thought to be the most reliable and repeatable gauges of sleep disturbance [34]. Habituation is yet another factor to consider, as there is clear evidence that this causes a variation in responses. There is also some disagreement among researchers regarding the noise descriptors that are most appropriate for this assessment.

Most studies trying to relate sleep disturbance to noise are associated with airport activities, while some studies are associated with road traffic noise. Recent public concern with wind farm noise has prompted several studies for these sources as well. Although L_{dn} is used most often to rate community annoyance, there is some debate about the relevance of using L_{dn} for sleep disturbance as a result of discrete events such as aircraft flyovers. Federal agencies in the US continue to support the use of L_{dn} as the descriptor of choice for airport noise evaluations, [35] but sleep disturbance criteria are also published in terms of the sound exposure level (SEL) of single aircraft flyovers. As mentioned earlier, SEL is based on the acoustic energy of a discrete event, such as an aircraft flyover or train pass-by.

The American National Standards Institute, in ANSI Standard S12.9-2008/Part 6, endorses two mathematical relationships for estimating awakenings caused by outdoor noise events heard in homes [36]: one for sleep disturbance from a new sound [37] and another for sleep disturbance from ongoing events based on a calculated probability of being awakened [38]. The mathematical relationships introduced in this standard are plotted in Figure 5.3. These curves are based on single aircraft flyover events causing behavioral awakenings, requiring the subject to be awake enough to push a button in response to the noise event. The curves in Figure 5.3 show the effects related to habituation to the sound source, for which the thresholds of awakening are significantly higher.

Another set of mathematically based dose–response curves related to awakening from noise uses descriptors related to sleep disturbance (Figure 5.4) [39]. As for annoyance ratings, sleep disturbance ratings are based on subjective levels from surveys, categorized as "highly sleep disturbed" for the ratings in the top 27% of the 0–100 disturbance scale. As for the annoyance curves, these have been categorized for different levels of self-reported sleep disturbance (with highly sleep-disturbed rating > 72 on the scale of 0–100, sleep-disturbed rating > 50 and little sleep-disturbed rating > 28) and by the major categories of transportation sources. Also consistent with the annoyance curves, aircraft noise was rated as the most disturbing to sleep while rail activities were rated the lowest of the transportation sources. This set of curves uses a different noise descriptor from the set in Figure 5.3, as these curves are based on a descriptor that averages noise levels from all transportation sources over an entire night (defined as between 11:00 pm and 7:00 am).

A recent Swedish study demonstrated the different conclusions that can be derived from different methods of data collection for sleep disturbance. A group of children and their parents were surveyed by questionnaire and actimetry to evaluate the effect of road traffic noise on sleep disturbance [40]. The results of this study showed better perceived sleep quality for the children than their parents, according to the questionnaire, while actimetry indicated the reverse for the same subjects.

There is a growing sentiment among researchers that the curves presented in Figures 5.3 and 5.4 are an over-simplification of the situation, and that there are many additional factors

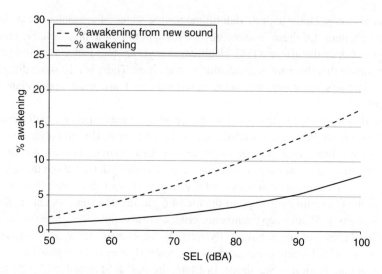

Figure 5.3 Dose–response curves for awakening from aircraft. SEL, sound exposure level. (Adapted from equations in ANSI S12.9-2008/Part 6 [36].)

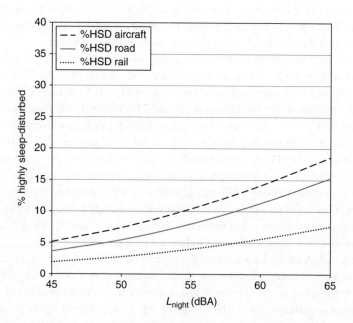

Figure 5.4 Dose–response curves for sleep disturbance from aircraft. %HSD, percentage of people highly sleep-disturbed by different forms of transportation. (Adapted from Miedema et al., [39] with permission of TNO.)

that contribute to the increase or decrease of noise-induced sleep disturbance beyond the outdoor noise level. Recent findings note that sleep is more likely to be disturbed in the latter part of the night and that most awakenings are from non-transportation sources [41]. In addition, interpretations and rating methods of awakenings vary widely between studies. Yet another factor is the building itself, which can significantly change interior sound levels. Therefore, basing awakening data on outdoor levels introduces a high degree of uncertainty.

One other method used for rating sleep problems associated with noise is tracking morning tiredness. A large 1991 study in the Netherlands (using close to 19,000 survey responses from subjects aged 15–74 years) showed a clear correlation between people who were tired in the morning with average night-time outdoor sound levels above 35 dBA from road traffic [42]. Confounding factors accounted for in the study included age, sex, body mass index, physical activity, employment, financial situation, smoking, alcohol use, education, and medications.

Although most published studies linking sleep disturbance with environmental noise in the past 50 years have been associated with transportation sources, a growing body of research is emerging looking at the links between industrial wind turbines and sleep disturbance. As for annoyance, sleep disturbance from wind turbines appears to begin at lower sound levels than that from transportation sources. This stems from the different frequency characteristics and modulation of levels associated with wind turbine noise. A 2010 US survey-based study of residents near wind farms in Maine showed increased sleep disturbance among residents within 1.4 km of wind turbines compared with those living farther away [43].

The results of these studies must be reviewed more closely, however, because there are more potential confounding factors when dealing with wind farms than there are with transportation sources. These include financial gain from the wind farm (from lease payments, reduced electricity rates, or increased tax revenues), fear of reduced property values, and fear of legal action to reduce the reported effects or attitude toward the wind farm and noise sensitivity to increase the reported effects. Wind turbine noise effects are discussed in more detail in Chapter 6.

5.5 Learning Disabilities

Since the 1970s, published results of studies have linked background noise to learning disabilities. A key determinant in this relationship is the restriction of communication ability as a result of noise. A common, simple descriptor used to rate communication ability in noise is the speech interference level (SIL), which is the arithmetic average of background SPLs in the octave bands between 500 and 4,000 Hz:

$$SIL = 1/4(SPL_{500} + SPL_{1,000} + SPL_{2,000} + SPL_{4,000}) \qquad (5.1)$$

Figure 5.5 shows the relationship between SIL and distance between listener and talker for "just-reliable communication" according to ANSI Standard S12.65 [44]. Different voice

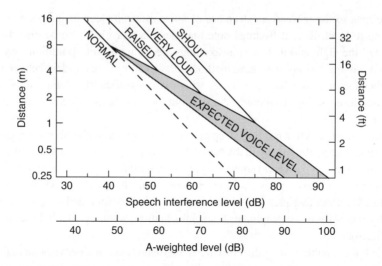

Figure 5.5 Speech interference with distance and sound pressure level for just-reliable communication [reprinted from ANSI/ASA S12.65-2006 (R 2011) *American National Standard for Rating Noise with Respect to Speech Interference*, © 2006, with the permission of the Acoustical Society of America, 1305 Walt Whitman Road, Suite 300, Melville, NY 11747]

levels required for different background sound levels are shown in this chart, with the shaded section indicating the expected voice levels associated with raising one's voice in response to elevated background noise levels. Note that A-weighted levels are listed as being roughly 8 dB higher than the SIL values in this chart, which is the case for most situations but may vary depending on the spectral content of the background sound. For example, if all SPL values in each octave band between 500 and 4,000 Hz are 50 dB, their A-weighted sum would be 56 dBA, which is 6 dB higher than the SIL of 50 dB. It is generally accepted that communication ability begins to be affected when background levels exceed 65 dBA.

Speech interference level is closely tied to speech intelligibility, which is a key parameter in effective learning. The information in Figure 5.5 assumes that the talker and listener are outdoors, so room effects are not associated with these values. As most classroom learning takes place indoors, room effects, such as reverberation, room resonance, and echoes, must also be considered for acceptable speech intelligibility.

These factors have been included in ANSI Standard S12.60 [45], which is used in the US by the Green Building Council in qualifications for acoustical performance for Leadership in Energy and Environmental Design (LEED) certifications of schools. Qualifications for the standard and this certification are divided into three categories for acoustics – reverberation, background noise, and sound insulation. The listed limits for learning spaces are 0.6–0.7 s for mid-frequency (500–2,000 Hz) reverberation time (RT_{60} – the time, in seconds, it takes for the SPL in a room to diminish by 60 dB after a sound source is deactivated) and background levels of 35 dBA and 55 dBC $L_{eq(1)}$ from all interior and exterior sources, with the goal of optimum teacher-to-student and student-to-student communications, although some studies have suggested lower limits (such as RT_{60} = 0.4–0.5 s and maximum background level of 30 dBA [46]). Finland has similar limits for background levels in classrooms, and even with a higher RT_{60} limit (0.9 s between 250 and 2,000 Hz), reverberation in classrooms is often too high for good speech intelligibility [47].

Students add absorption to a room and, as RT_{60} decreases with increasing room absorption, unoccupied classrooms tend to have higher RT_{60} values than occupied classrooms. They also obviously add noise to classrooms, so all specifications are based on unoccupied spaces.

Another factor relating to ideal communications in classrooms is the signal-to-noise ratio or speech–noise level difference (denoted S/N). S/N is the arithmetic difference between the signal (speech in this case) and the background noise level. Studies have shown that the ideal S/N for learning is 15–20 dBA (for 95% speech intelligibility), with the higher values relevant to younger students [48]. However, S/N must be tied to RT_{60} for the most meaningful results. A recent study on elementary school children showed that with constant S/N, speech intelligibility increased with decreasing RT_{60}, but intelligibility was maximized in an RT_{60} range of 0.3–0.9 s [49]. Speech intelligibility dropped for RT_{60} values less than 0.3 s, showing the value of some reverberation or early reflections to enhance the sound in the classroom.

In terms of the general acoustical environment in learning spaces, a recent Italian study showed that students were more disturbed by intermittent acoustic events than by constant noises [50]. A UK study surveyed over 2,000 primary school children (6–11 years old) and teachers, showing that external transportation sources were the most annoying and detrimental to learning [51]. Another UK study rated performance with noise exposures and showed that both exterior and interior noise sources have a negative effect, with older students (11 years of age) being more affected than younger ones. This study showed that, in general, students in classrooms in which background levels (L_{90}) exceeded 50 dBA failed to meet government literacy targets [52].

The subject of speech intelligibility in classrooms is usually limited to background noise and reverberation since low background noise levels and low reverberation times are assumed for those spaces. Some research studies have used more complex speech intelligibility descriptors, most notably articulation index (AI) and speech intelligibility index (SII). These parameters range from 0 to 1, with 0 representing no intelligibility and 1 representing total intelligibility. High speech intelligibility implies low speech privacy, and low speech intelligibility implies high speech privacy. These parameters are used more often for speech privacy ratings in spaces with higher background levels and reverberation times and, as such, are discussed in more detail in the speech privacy section of Chapter 8.

5.5.1 Cognitive Development/School Performance

Three categories of cognitive performance have been studied the most with regard to noise effects – reading comprehension, memory, and standardized test scores. Many studies linking noise with reading comprehension have been performed since the 1970s, and the majority of these have been based on transportation sources [53]. One set of studies in a New York City elementary school revealed a clear correlation. In this case, an elevated rail line passed on one side of the school building and students in classrooms on the side facing the rail line were lagging in reading level test scores by 3–11 months compared with students in classrooms on the side of the building not facing the rail line [54]. Passing trains generated sound levels of up to 89 dBA in the classrooms every 4.5 mins for up to 30 s with each pass-by, thereby disrupting classroom activities. Noise control measures implemented on the rail line and in the classrooms lowered the train noise in the classrooms by up to 8 dBA and resulted in reading test scores being the same for students in classrooms on both sides of the building [55].

Although most studies in this category use scores on standardized tests as the basis for their conclusions, some have attempted to delve deeper into the reasons behind the variations in test scores as they are associated with noise exposures. Another New York City study on first- and second-grade school children was performed to determine whether elevated noise levels (outdoor levels > 65 dBA L_{dn} in this case, mainly from aircraft noise) affected language development as the root cause of reading disabilities [56]. This study confirmed that children exposed to high levels of background noise have poorer reading skills than those in quieter areas, and also confirmed that high noise exposure impairs speech perception, but did not confirm the hypothesis that noise impairs language development.

In an example of the effects of noise on memory in a learning environment, a recent study on fourth-grade school children revealed that elevated noise level exposures can significantly affect intentional, but less so incidental, memory [57]. Intentional memory is related to tasks for which the subjects are told that they will be asked about specific material, and incidental memory is related to tasks for which the subjects are not told they will be asked about the material afterwards. This study was performed in small towns and villages in an area outside Innsbruck, Austria, where there are areas exposed to elevated rail and road traffic noise (>60 dBA L_{dn}) that can be compared with areas having lower levels of background noise (<50 dBA L_{dn}).

Another clear example comes from a study conveniently performed in Germany before Munich International Airport was closed at one location and moved to another [58]. Elementary school children (aged 8–12) were tested for various cognitive tasks (such as reading, memory, attention, and speech perception) at these locations and there were clear indications of memory and reading impairments associated with the opening and closing of the airports. Learning performance improved near the location of the old airport after it closed and the same performance degraded near the new airport after it opened.

Several large studies have been performed recently to provide more definitive answers to the questions regarding the effects of noise on learning. One of these, the RANCH project (Road traffic and Aircraft Noise exposure and Children's cognitive Health), was performed between 2001 and 2003. This involved 2,010 children aged 9–10 from 89 schools near three European airports – Amsterdam Schiphol in the Netherlands, Madrid Barajas in Spain, and London Heathrow in the UK. The goal of the study was to determine the relationship between aircraft and road traffic noise on cognitive performance and health. The results of the study showed a clear relationship between aircraft noise at home and school, and between aircraft noise and reading comprehension using standardized tests. The study did not reveal such relationships for traffic noise, most likely because aircraft noise is more intermittent than traffic noise, thus being more effective at disrupting concentration [59,60]. Outdoor 16-hour (7:00 am to 11:00 pm) L_{eq} values ranged from 30 to 70 dBA and the confounding factors accounted for were mother's education, socioeconomic status, parental support for schoolwork, prolonged illness, and classroom noise insulation.

Another of these studies was sponsored by the US Federal Interagency Committee on Aviation Noise (FICAN) in 2000 to determine the effects of airport closures and sound insulation in schools near airports related to cognitive performance. This study included 35 schools (elementary to high schools) near three airports in the US and the results were based on changes in standardized test scores. Typical sound insulation programs for buildings near airports include replacing exterior windows and doors and adding roof insulation, totaling a maximum of 5–7 dBA of new overall noise reductions to classrooms. The results of this study showed reductions in failure rates for high school students but not for elementary or middle-school students when noise levels were decreased by such means [61].

Study results published in 2014 by the Transportation Research Board in the US confirmed the role of building insulation with regard to improving standardized test scores from a much larger student sample than used in the FICAN study. This study evaluated test scores from 6,198 schools near 46 airports around the US [62]. Rather than using absolute outdoor sound levels as most other of these types of studies have done, this study showed more consistent results when comparing aircraft noise levels with the background levels without aircraft flyovers. Looking at before and after sound insulation in the same schools in the study, the insulation effectively cancelled out the negative effects of aircraft noise exposure on test scores.

The latest European study on the cognitive effects of noise, the NORAH study (Noise-Related Annoyance, cognition, and Health), was performed in 2012 in Germany. This study involved 85 second-grade classes from 29 primary schools for 1,243 children living near Frankfurt Airport. The results yielded a 1-month reading delay associated with a 10 dBA L_{eq} (during daytime hours) increase in aircraft noise both at home and at school [63].

The World Health Organization released a summary of noise effects on health and cognition in 2011, and Figure 5.6 provides a summary from that document showing the average trend of the percentage of students affected (in terms of reading comprehension, memory, and standardized test scores) by outdoor average noise level exposures. The key points to bear in mind with this are that cognitive effects are minimal below outdoor levels of 50 dBA L_{dn} and the curve in Figure 5.6 is an average and is by no means absolute.

5.5.2   Office/Occupational

Work environments with potentially hazardous noise levels are discussed in Chapters 4 and 8, so this discussion relates to exposure levels below hearing damage limits. In typical office environments, noise is one of the most common elements affecting worker productivity, especially with the prevalence of the open-plan office design. Many studies have been performed since the 1960s relating noise to productivity and speech privacy. Studies involving task performance in the presence of noise have had mixed results, based on the level and type of noise being evaluated [64]. Speech privacy is a much larger concern than background noise level, and it has been shown that occupants of open-plan office environments without cubicles are more satisfied than those in cubicles, presumably because those in cubicles

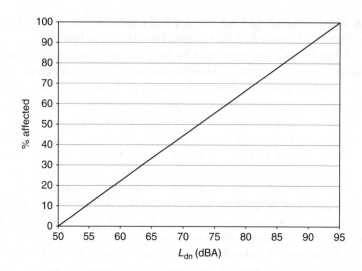

Figure 5.6 Exposure-risk curve for people cognitively affected by noise, in terms of outdoor levels (from WHO 2011 [65], with permission of WHO)

cannot see whether anyone nearby is listening to their conversations [66]. It has also been shown that irrelevant speech can be very disruptive, reducing productivity by a factor of one-third of the level in a quiet environment [67], and low-level office noise (an average of 55 dBA with peaks up to 65 dBA) can increase stress hormonal concentrations without workers being aware of it [68]. In addition to the sound level, other aspects of the office noise environment that add to annoyance and distraction are the degree of control the worker has over the noise source, the self-rated necessity of the noise source, personal noise sensitivity, and the predictability of the noise [69].

The simplest, most effective solution to stress and distractions caused by open-plan arrangements is an effective sound masking system [70]. An effective sound masking system raises the background sound level in an office environment without most people noticing the sound has been added. Masking systems typically generate red (or brown) noise rather than the expected white or pink noise, as red noise mimics heating, ventilation and air conditioning noise more closely than white or pink noise, with a 6 dB drop in level with increasing octave bands. Masking system loudspeakers are usually positioned in ceiling areas of open-plan offices to spread the sound evenly throughout the space; however, individual adjustable units are available for each desk or cubicle. Individual units minimize the potential for complaints from those who are sensitive to the masking sounds, but the optimum SPL for masking systems is in the 45–48 dBA range. This is high enough in level to effectively mask most distracting office sounds while not being so loud as to be distracting by itself.

The effects of outdoor noise on worker satisfaction were addressed in a recent polling of more than 23,000 office workers in four countries (US, Finland, Canada, and Australia), showing that, for even higher levels of exposure, workers near operable windows are more satisfied than those near sealed windows or those away from any windows [71]. The study also showed that complaints about indoor noise sources outnumbered complaints about outdoor noise sources by a factor of 10, with the highest percentage of complaints related to co-workers speaking with each other and telephone conversations. Of outdoor noise sources that were concerns, construction noise was rated the highest.

There is an emerging body of research on the relationships between noise and occupational performance in the medical community. The current emphasis is on the topics of alarm fatigue and operation room noise. Alarm fatigue is a term used to describe a growing problem in hospitals whereby alarms are either ignored or deactivated. There are many reasons for this, most notably the high false alarm rate (as high as 85%), which leads to the desensitization of hospital staff members to audible alarms [72]. There is also limited capacity for prioritizing alarms, potentially leading to fatalities [73]. Nurses' responses to alarms are based on workload, their perceived probability of the alarm being false, and each patient's condition [74].

Also adding to alarm fatigue and diminished performance in hospitals are high noise levels in general. Studies in hospitals worldwide have shown a trend of increased noise levels over the past 50 years, with L_{eq} values increasing from 57 dBA in 1960 to 72 dBA in 2005 during the day and from 42 dBA in 1960 to 60 dBA in 2005 during night-time hours [75]. Noise levels in operating rooms can be particularly loud, with average levels of up to 70 dBA and peak levels of more than 120 dB (unweighted) [76], raising a concern not only about clear speech communication but also about the potential for hearing loss over time. While hospitals can be stressful workplaces for other reasons, noise adds to psychological stress in these environments. A 2011 European study of noise from 156 operations in a pediatric operating room demonstrated that a 4 dBA reduction in mean SPL resulted in a 20% reduction in surgeons' stress hormone concentrations and a 50% reduction in postoperative complications from a conscious reduction of noise in the operating room [77]. These noise reductions were entirely a result of behavioral changes supported by briefings and warning signals that changed colors when specified limits were exceeded. A recent survey of Swedish nurses also showed not only that they felt irritation and fatigue as a result of the sound environment in hospitals, but that they overwhelmingly thought that acutely ill patients can develop what is known as intensive care unit syndrome from their noise exposures [78]. Intensive care unit syndrome is characterized by symptoms of distress, instability, vulnerability, and fear, accompanied by hallucinations. This is not brought on entirely by noise, but noise is felt to be a significant contributor to its onset.

A recent Polish study added to the literature regarding the annoyance of low-frequency noise (defined in this study as a measured difference between dBC and dBA levels greater than 15), but in occupational environments. In this case, 276 workers rated moderate noise (averaging 58 dBA) having dominant low-frequency components as being more annoying than background sounds of a similar level without dominant low-frequency components [79]. This study also concluded that A-weighted levels associated with sounds dominated by low-frequency noise adequately correlate with annoyance ratings, in opposition to conclusions in many other low-frequency noise annoyance studies.

Several studies have taken into account the complexity of jobs when considering noise effects. These have shown that occupational noise higher than an average of 80 dBA has a negative effect on job satisfaction for those performing complex job tasks, but moderate noise can be beneficial for those working in more simple jobs [80]. The CORDIS (Cardiovascular Occupational Risk factors Determination in Israel) study, performed from

1985 to 1987 on 880 white-collar workers from 21 organizations, yielded results showing a correlation between sickness absence for female employees only working in complex jobs in high noise environments compared with low noise environments, but the absolute noise levels were not clearly conveyed in the literature [81].

A clear relationship between occupational noise and job satisfaction was illustrated by a 1997 Italian study in which a group of workers were moved to a facility where noise levels were roughly 10 dBA less than those in their former facility (in the upper 70s to mid-80s on the dBA scale). Increased job satisfaction, reduction of self-reported stress, and positive attachments to the company all resulted from changing this one measure of noise control in the working environment [82].

5.6   Emotional Effects

Most of the research relating noise and emotional effects deals with the topics of aggression and general mental performance. When addressing aggressive behaviors, a key aspect of the environment is the feeling of control the subject has over the noise source, along with its predictability. If one doesn't feel a sense of control over the noise source, the next logical step is to attempt to gain that control. Depending on the circumstance, aggression is one option for gaining that control. Although seemingly intuitive, this has been verified by experiments since the 1970s involving subjects inflicting varying intensities of electric shocks on another subject while being exposed to varying levels of noise [83,84]. This is especially the case when the subject has a perceived lack of control over the noise source.

The idea of learned helplessness is tied to the condition of uncontrollable noise exposures. As mentioned earlier in this chapter, this can illicit increases in cortisol production and thus stress reactions. It has been shown that this condition can also affect memory, especially in the presence of uncontrollable noise. A British experiment exposing subjects to uncontrollable loud spoken noises increased tension and the recall of negative trait words from a list of positive and negative ones, while not affecting those with the ability to control the noise sources. The study summary goes on to state that the negative memory bias caused by the experiment is similar to that found in clinically depressed patients [85].

The largest complicating factor in evaluating emotional effects of noise is individual noise sensitivity. This is based on personality type and experiences and, while personality type can be used in generalizations, personal traumas and experiences cannot. A 12-year overview on these issues concluded that those with stable personalities, extroversive tendencies, and low noise sensitivities tend to adapt better mentally to noise than those with the opposite traits [86]. This premise has been supported by recent studies, but with regard to mental performance, a recent Iranian study showed that moderate levels (71 dBA L_{eq}) of traffic noise improved attention and concentration compared with a quiet setting, especially for extroverts and males. Introverts and females had no negative effects except for a reduction in the pace of work [87].

A study on 1,403 children aged 8–11 years in Austrian villages exposed to low (<50 dBA L_{dn}) and high (>60 dBA L_{dn}) environmental noise showed small links between higher noise and mental health (self-reported anxiety and depression) but more significant links for children who had low birth weights [88]. Another larger study of more than 2,000 children aged 9–10 (the RANCH study, mentioned earlier in the discussion on cognitive effects) yielded no clear links between environmental noise exposures and mental health (in terms of psychological distress) [89]. The only effects of note from that study were for exacerbated symptoms for hyperactive children, but this has not been supported elsewhere.

For adults, there is little evidence linking general environmental noise with psychiatric disorders. This was confirmed as part of a large UK road traffic noise study done for more than 2,000 men aged 50–64 in Caerphilly, Wales [90]. Only links to annoyance and anxiety could be implied from this study. Subjective noise sensitivity plays a major role in these findings [91]. The idea of noise sensitivity also makes a difference in typical coping mechanisms. A study of nearly 2,000 residents, aged 25–65, in rural areas of the Austrian Alps where road traffic noise (especially from trucks) has tripled over the past 40 years yielded significant associations between $L_{eq(24)}$ exposures > 55 dBA and annoyance accompanied by taking actions to control the noise, such as closing and replacing windows, moving noise-sensitive rooms within houses, filing complaints, and joining community groups to address the noise. However, those considering themselves as noise-sensitive took fewer actions to control the noise and had more health issues [92].

References

[1] Halpern, D.L., Blake, R. and Hillenbrand, J. (1986). "Psychoacoustics of a chilling sound." *Perception and Psychophysics*, 39(2): 77–80.

[2] Fields, J.M. (1993). "Effect of personal and situational variables on noise annoyance in residential areas." *Journal of the Acoustical Society of America*, 93(5): 2753–2763.

[3] Flindell, I.H. and Stallen, P.J.M. (1999). "Non-acoustical factors in environmental noise." *Noise & Health*, 3: 11–16.

[4] Guski, R., Felscher-Suhr, U. and Schuemer, R. (1999). "The concept of noise annoyance: how international experts see it." *Journal of Sound and Vibration*, 223(4): 513–527.

[5] Leventhall, H.G. (2004). "Low frequency noise and annoyance." *Noise & Health*, 6(23): 59–72.

[6] Møller, H. (1987). "Annoyance of audible infrasound." *Journal of Low Frequency Noise and Vibration*, 6(1): 1–17.

[7] Andersen, J. and Møller, H. (1984). "Equal annoyance contours for infrasonic frequencies." *Journal of Low Frequency Noise and Vibration*, 3(3): 1–8.

[8] Pawlaczyk-Luszczynska, M., et al. (2003). "Assessment of annoyance from low frequency and broadband noises." *International Journal of Occupational Medicine and Environmental Health*, 16(4): 337–343.

[9] Kjellberg, A., Goldstein, M. and Gamberale, F. (1984). "An assessment of dB(A) for predicting loudness and annoyance of noise containing low frequency components." *Journal of Low Frequency Noise and Vibration*, 3(3): 10–16.

[10] Persson, K. and Björkman, M. (1988). "Annoyance due to low frequency noise and the use of the dB(A) scale." *Journal of Sound and Vibration*, 127(3): 491–497.

[11] Schultz, T.J. (1978). "Synthesis of social surveys on noise annoyance." *Journal of the Acoustical Society of America*, 64(2): 377–405.

[12] Fidell, S, Barber, D.S. and Schultz, T.J. (1991). "Updating a dosage-effect relationship for the prevalence of annoyance due to general transportation noise." *Journal of the Acoustical Society of America*, 89(1): 221–233.

[13] Miedema, H.M.E. and Vos, H. (1998). "Exposure-response relationships for transportation noise." *Journal of the Acoustical Society of America*, 104(6): 3432–3445.

[14] Miedema, H.M.E and Oudshoorn, C.G.M. (2001). "Annoyance from transportation noise: Relationships with exposure metrics DNL and DENL and their confidence intervals." *Environmental Health Perspectives*, 109(4): 409–416.

[15] Kim, J., et al. (2010). "Noise-induced annoyance from transportation noise: Short-term responses to a single noise source in a laboratory." *Journal of the Acoustical Society of America*, 127(2): 804–814.

[16] Lim, C., et al. (2006). "The relationship between railway noise and community annoyance in Korea." *Journal of the Acoustical Society of America*, 120(4): 2037–2042.

[17] Öhrström, E., et al. (2007). "Annoyance due to single and combined sound exposure from railway and road traffic." *Journal of the Acoustical Society of America*, 122(5): 2642–2652.

[18] van Kempen, E.E.M.M., et al. (2009). "Children's annoyance reactions to aircraft and road traffic noise." *Journal of the Acoustical Society of America*, 125(2): 895–904.

[19] Van Gerven, P.W.M., et al. (2009). "Annoyance from environmental noise across the lifespan." *Journal of the Acoustical Society of America*, 126(1): 187–194.

[20] Miedema, H.M.E. and Vos, H. (1999). "Demographic and attitudinal factors that modify annoyance from transportation noise." *Journal of the Acoustical Society of America*, 105(6): 3336–3344.

[21] Miedema, H.M.E., Fields, J.M. and Vos, H. (2005). "Effect of season and meteorological conditions on community noise annoyance." *Journal of the Acoustical Society of America*, 117(5): 2853–2865.

[22] Job, R.F.S. (1988). "Community response to noise: A review of factors influencing the relationship between noise exposure and reaction." *Journal of the Acoustical Society of America*, 83(3): 991–1001.

[23] Wenjing, N. and Hui, M. (2014). "Influence of color and brightness condition on annoyance evaluation caused by road traffic noise in indoor environment." *Proceedings of* ICBEN *2014*, Nara, Japan.

[24] Janssen, S.A., et al. (2011). "A comparison between exposure-response relationships for wind turbine annoyance and annoyance due to other noise sources." *Journal of the Acoustical Society of America*, 130(6): 3746–3753.

[25] Maynard, R. *Environmental Noise and Health in the UK*. Oxfordshire: Health Protection Agency, 2010.

[26] Ising, H. and Braun, C. (2000). "Acute and chronic endocrine effects of noise: Review of the research conducted at the Institute for Water, Soil and Air Hygiene." *Noise & Health*, 7:7–24.

[27] Melamed S. and Bruhis S. (1996). "The effects of chronic industrial noise exposure on urinary cortisol, fatigue and irritability – A controlled field experiment." *Journal of Occupational and Environmental Medicine*, 38(3): 252–256.

[28] Ljungberg, J. K. and Neely, G. (2007). "Stress, subjective experience and cognitive performance during exposure to noise and vibration." *Journal of Environmental Psychology*, 27(1): 44–54.

[29] Evans, G.W., et al. (2001). "Community noise exposure and stress in children." *Journal of the Acoustical Society of America*, 109(3): 1023–1027.

[30] Ising, H. and Ising, M. (2002). "Chronic cortisol increases in the first half of the night caused by road traffic noise." *Noise & Health*, 4(16): 13–21.

[31] Persson Waye, K., et al. (2002). "Low frequency noise enhances cortisol among noise sensitive subjects during work performance." *Life Sciences*, 70: 745–758.

[32] Haines, M., et al. (2001). "A follow-up study of effects of chronic aircraft noise exposure on child stress responses and cognition." *International Journal of Epidemiology*, 30:839–845.

[33] Babisch, W., et al. (2001). "Increased catecholamine levels in urine in subjects exposed to road traffic noise. The role of stress hormones in noise research." *Environment International*, 26: 475–481.

[34] Basner, M., Samel, A. and Isermann, U. (2006). "Aircraft noise effects on sleep: Application of the results of a large polysomnographic filed study." *Journal of the Acoustical Society of America*, 119(5): 2772–2784.

[35] Finegold, L.S., Harris, S. and von Gierke, H.E. (1994). "Community annoyance and sleep disturbance: updated criteria for assessing the impacts of general transportation noise on people." *Noise Control Engineering Journal*, 42(1): 25–30.

[36] Acoustical Society of America. *ANSI S12.9-2008/Part 6. Quantities and Procedures for Description and Measurement of Environmental Sound – Part 6: Methods for estimation of awakenings associated with outdoor noise events heard in homes*. New York: American Institute of Physics, 2008.

[37] Federal Interagency Committee on Aviation Noise. *Effects of Aviation Noise on Awakenings from Sleep*. Washington, DC: FICAN, 1997.

[38] Anderson, G.S. and Miller, N.P. (2007). "Alternative analysis of sleep-awakening data." *Noise Control Engineering Journal*, 55(2): 224–245.

[39] Miedema, H.M.E, Passchier-Vermeer, W. and Vos, H. *Elements for a Position Paper on Night-time Transportation Noise and Sleep Disturbance*. TNO Inro Report 2002-59. Delft: TNO, 2002.

[40] Öhrström, E., et al. (2006). "Effects of road traffic noise on sleep: Studies on children and adults." *Journal of Environmental Psychology*, 26: 116–126.

[41] Michaud, D.S., et al. (2007). "Review of field studies of aircraft noise-induced sleep disturbance." *Journal of the Acoustical Society of America*, 121(1): 32–41.

[42] de Kluizenaar, Y., et al. (2009). "Long-term road traffic noise exposure is associated with an increase in morning tiredness." *Journal of the Acoustical Society of America*, 126(2): 626–633.

[43] Nissenbaum, M.A., Aramini, J.J. and Hanning, C.D. (2012). "Effects of industrial wind turbine noise on sleep and health." *Noise & Health, Sep–Oct*; 14(60): 237–243.

[44] Acoustical Society of America. *ANSI S12.65-2006 (R2011). For Rating Noise with Respect to Speech Interference*. New York: American Institute of Physics, 2011.

[45] Acoustical Society of America. ANSI S12.60-2010/Part 1. *Acoustical Performance Criteria, Design Requirements, and Guidelines for Schools, Part 1: Permanent schools*. New York: American Institute of Physics, 2010.

[46] Bradley, J.S. (1986). "Speech intelligibility studies in classrooms." *Journal of the Acoustical Society of America*, 80(3): 846–854.

[47] Pekkarinen, E., Viljanen, V. (1991). "Acoustic conditions for speech communication in classrooms." *Journal of Scandinavian Audiology*, 20: 257–263.

[48] Bradley, J.S. and Sato, H. (2008). "The intelligibility of speech in elementary school classrooms." *Journal of the Acoustical Society of America*, 123(4): 2078–2086.

[49] Yang, W. and Bradley, J. S. (2009). "Effects of room acoustics on the intelligibility of speech in classrooms for young children." *Journal of the Acoustical Society of America*, 125(2): 922–933.

[50] Astolfi, A. and Pellerey, F. (2008). "Subjective and objective assessment of acoustical and overall environmental quality in secondary classrooms." *Journal of the Acoustical Society of America*, 123(1): 163–173.

[51] Dockrell, J.E. and Shield, B. (2004). "Children's perceptions of their acoustic environment at school and at home." *Journal of the Acoustical Society of America*, 115(6): 2964–2973.

[52] Shield, B.M. and Dockrell, J.E. (2008). "The effects of environmental and classroom noise on the academic attainments of primary school children." *Journal of the Acoustical Society of America*, 123(1): 133–144.

[53] Clark, C. and Stansfeld, S.A. (2007). "The effect of transportation noise on health and cognitive development: a review of recent evidence." *International Journal of Comparative Psychology*, 20: 145–158.

[54] Bronzaft, A.I. and McCarthy, D.P. (1975). "The effect of elevated train noise on reading ability." *Environment and Behavior*, 7(4), pp. 517–528.

[55] Bronzaft, A. (1981). "The effect of a noise abatement program on reading ability." *Journal of Environmental Psychology*, 1: 215–222.

[56] Evans, G.W. and Maxwell, L. (1997). "Chronic noise exposure and reading deficits: The mediating effects of language acquisition." *Environment and Behavior*, 29(5): 638–656.

[57] Lercher, P., Evans, G.W. and Meis, M. (2003). "Ambient noise and cognitive processes among primary schoolchildren." *Environment and Behavior*, 35: 725–735.

[58] Hygge, S. Evans, G.W. and Bullinger, M. (2002). "A prospective study of some effects of aircraft noise on cognitive performance in school children." *Psychological Science*, 13(5): 469–474.

[59] Stansfeld, S, et al. (2005). "Aircraft and road traffic noise and children's cognition and health: a cross-national study." *The Lancet*, 365: 1942–1949.

[60] Clark, C., et al. (2006). "Exposure-effect relations between aircraft and road traffic noise exposure at school and reading comprehension: The RANCH project." *American Journal of Epidemiology*, 163(1): 27–37.

[61] Federal Interagency Committee on Aviation Noise. *Findings of the FICAN Pilot Study on the Relationship between Aircraft Noise Reduction and Changes in Standardized Test Scores*. Washington, DC: FICAN, 2007.

[62] Sharp, B.H., et al. *Assessing Aircraft Noise Conditions Affecting Student Learning. Final Report for ACRP Project 02-26*. Transportation Research Board of the National Academies, 2014.

[63] Klatte, M., et al. *Effects of Chronic Flight Noise Exposure on Cognitive Performance and Quality of Life in Primary School Children (in German)*. Kelsterbach, Germany: Non-profit Umwelthaus GmbH, 2014.

[64] Theologus, G.C., Wheaton, G.R. and Fleishman, E.A. (1974). "Effects of intermittent, moderate intensity noise stress on human performance." *Journal of Applied Psychology*, 59 (5): 539–547.

[65] World Health Organization. *Burden of Disease from Environmental Noise: Quantification of Healthy Life Years Lost in Europe*. Copenhagen: WHO, 2011.

[66] Jensen, K. and Edward, A. (2005). "Acoustical quality in office workstations, as assessed by occupant surveys." *Proceedings of Indoor Air 2005*, Beijing, China, 2401–2405.

[67] Banbury, S. and Berry, D.C. (1998). "Disruption of office-related tasks by speech and office noise," *British Journal of Psychology*, 89(3): 499–517.

[68] Evans, G.W. and Johnson, D. (2000). "Stress and open-office noise." *Journal of Applied Psychology*, 85(5): 779–783.

[69] Kjellberg, A., et al. (1996). "The effects of nonphysical noise characteristics, ongoing task and noise sensitivity on annoyance and distraction due to noise at work." *Journal of Environmental Psychology*, 16: 123–126.

[70] Loewen, L.J. and Suedfeld, P. (1992). "Cognitive and arousal effects of masking office noise." *Environment and Behavior*, 24(3): 381–395.

[71] Goins, J, Chun, C. and Zhang, H. "User perspectives on outdoor noise in buildings with operable windows." UC Berkeley Center for the Built Environment, 2012.

[72] Barach, P., et al. (2014). "Redesigning hospital alarms for patient safety: alarmed and potentially dangerous." *Proceedings of ICBEN 2014*, Nara, Japan.

[73] Solet, J.M. and Barach, P.R. (2012). "Managing alarm fatigue in cardiac care." *Progress in Pediatric Cardiology*, 33: 85–90.

[74] Cvach, M. (2012). "Monitor alarm fatigue: an integrative review." *Biomedical Instrumentation & Technology*, 46(4): 268–277.

[75] Busch-Vishniac, et al. (2005). "Noise levels in Johns Hopkins Hospital." *Journal of the Acoustical Society of America*, 118(6): 3629–3645.

[76] Kracht, J.M., Busch-Vishniac, I.J. and West, J.E. (2007). "Noise in the operating rooms of Johns Hopkins Hospital." *Journal of the Acoustical Society of America*, 121(5): 2673–2680.

[77] Engelmann, C.R., et al. (2013). "A noise-reduction program in a pediatric operation theatre is associated with surgeon's benefits and a reduced rate of complications, a prospective controlled clinical trial." *Annals of Surgery*, 00(00): 1–9.

[78] Ryherd, E.E., Waye, K.P. and Ljungkvist, L. (2008). "Characterizing noise and perceived work environment in a neurological intensive care unit." *Journal of the Acoustical Society of America*, 123(2): 747–756.

[79] Pawlaczyk-Luszczynska, M., et al. (2009). "Annoyance related to low frequency noise in subjective assessment of workers." *Journal of Low Frequency Noise, Vibration and Active Control*, 28(1): 1–17.

[80] Melamed, S., Fried, Y., and Froom, P. (2001). "The interactive effect of chronic exposure to noise and job complexity on changes in blood pressure and job satisfaction: a longitudinal study of industrial employees." *Journal of Occupational Health Psychology*, 6(3):182–195.

[81] Fried, Y., Melamed, S., and Ben-David, H.A. (2002). "The joint effects of noise, job complexity, and gender on employee sickness absence: An exploratory study across 21 organizations – the CORDIS study." *Journal of Occupational and Organizational Psychology*, 75 (2): 131–144.

[82.] Raffaello, M. and Maass, A. (2002). "Chronic exposure to noise in industry: The effects on satisfaction, stress symptoms, and company attachment." *Environment and Behavior*, 34: 651–671.

[83.] Donnerstein, E. and Wilson, D.W. (1976). "Effects of noise and perceived control on ongoing and subsequent aggressive behavior." *Journal of Personality and Social Psychology*, 34(5): 774–781.

[84] Geen, R.G. and McCown, E.J. (1984). "Effects of noise and attack on aggression and physiological arousal." *Motivation and Emotion*, 8(3): 231–241.

[85] Willner, P. and Neiva, J. (1986). "Brief exposure to uncontrollable but not to controllable noise biases the retrieval of information from memory." *British Journal of Clinical Psychology*, 25: 93–100.

[86] Belojevic, G., Jakovljevic, B. and Slepcevic, V. (2003). "Noise and mental performance: Personality attributes and noise sensitivity." *Noise & Health*, 6(21): 77–89.

[87] Alimohammadi, I., et al. (2013). "The effects of road traffic noise on mental performance." *Iranian Journal of Environmental Health Science & Engineering*, 10(18).

[88] Lercher, P., et al. (2002). Ambient neighborhood noise and children's mental health." *Occupational and Environmental Medicine*, 59(6): 380–386.

[89] Stansfeld, S.A., et al. (2009). "Aircraft and road traffic noise exposure and children's mental health." *Journal of Environmental Psychology*, 29: 203–207.

[90] Stansfeld, S., et al. (1996). "Road traffic noise and psychiatric disorder: prospective findings from the Caerphilly study." *British Medical Journal*, 313: 266 – 267.

[91] Stansfeld, S.A., et al. (1993). "Road traffic noise, noise sensitivity and psychological disorder." *Psychological Medicine*, 23: 977–985.

[92] Lercher, P. and Kofler, W.W. (1996). "Behavioral and health responses associated with road traffic noise along Alpine through-traffic-routes." *The Science of the Total Environment*, 189/190: 85–89.

6

Sound Sources Associated with Negative Effects

6.1 Introduction

As long as human civilizations have existed, anthropogenic sound sources have contributed to the acoustic environment. The explosion of technological advances in the past century has increased the exposure of these sound sources to nearly everyone on the planet. These sources provide conveniences for humanity that have never been experienced in recorded history, but some of the consequences of such convenience can be the effects described in the last two chapters. Artificial sound sources affecting most people can be divided into three categories – transportation, utilities, and recreational. This chapter summarizes the acoustic qualities of these sources to correlate with their most common effects. The topic of hums (or phantom sounds) is included in this discussion but in a separate section as many of their sources are unknown. The chapter ends with a brief discussion of how the negative physiological effects of sound are being used in non-lethal weapons, through both real examples and folklore.

6.2 Transportation Sources

It is very difficult to avoid the sounds associated with transportation sources in the 21st century. The convenience of worldwide travel has made these types of sources the most common that people are exposed to on a daily basis. Figure 6.1 shows a map of the continental US marked with transportation noise source paths to illustrate the remaining quiet areas available as of 2007. The advent of air and road travel in the 20th century has been the main cause of this widespread increase in environmental noise exposures. Natural quiet is defined here as being removed enough from transportation sources that they do not dominate the background sound environment. Using this criterion, natural quiet could be

The Effects of Sound on People, First Edition. James P. Cowan.
© 2016 John Wiley & Sons, Ltd. Published 2016 by John Wiley & Sons, Ltd.

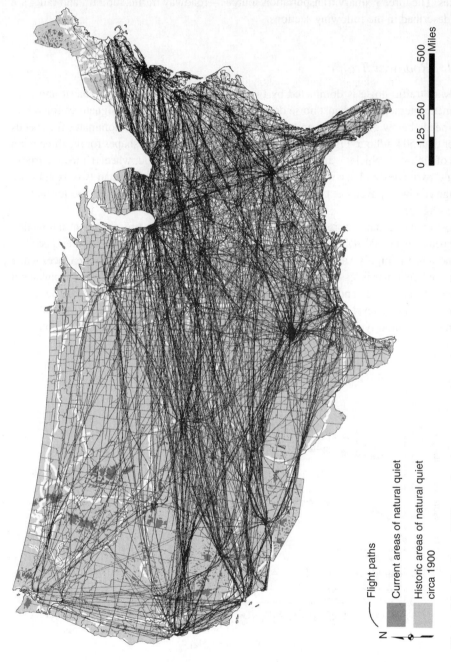

Figure 6.1 Continental US map showing transportation noise coverage to compare areas of natural quiet in 1900 with those in 2007 (from data compiled by and reproduced with permission of the Noise Pollution Clearinghouse)

Flight paths

Current areas of natural quiet

Historic areas of natural quiet circa 1900

N

0   125   250   500
                    Miles

found in only 2.6 % of the continental US in 2007, compared with 75% so defined before the aviation and automobile industries began, when trains were the only transportation sources. The three primary transportation sources – roadway traffic, aircraft, and rail – are each described in the following sections.

6.2.1 Roadway Traffic

Roadway traffic noise is dominated by two primary vehicle sources – exhaust and tire–pavement interaction. Exhaust noise dominates roadway traffic noise signatures for vehicular speeds below 50 km/hour and tire–pavement interaction noise dominates for speeds higher than 50 km/hour. Figures 6.2 and 6.3 show typical spectral shapes for the three main types of highway vehicles – automobiles (having two axles and four wheels), medium trucks (having two axles and six wheels), and heavy trucks (having more than two axles) – on average roadway pavement types. Figure 6.4 shows the trend of traffic noise levels with increasing speed.

Most engine exhaust noise is dominant below 250 Hz while tire–pavement interaction noise peaks in the 800–1,200 Hz range, increasing in level with increasing vehicle speed [2]. Engine noise is highly variable, depending on engine type, age, and maintenance, while tire–pavement noise is less variable but still significantly variable due to different pavement surface characteristics and different tire designs and levels of wear. Tires make less of a difference in noise level (up to 3 dBA) compared with pavement types (up to 10 dBA) [3]. The most common types of roadway surface materials are dense-graded asphalt concrete,

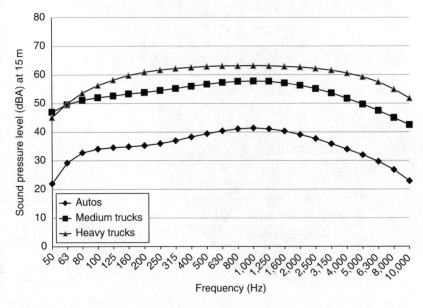

Figure 6.2 Typical roadway traffic noise spectra at 16 km/hour speeds on average pavements (adapted from data in the Traffic Noise Model manual [1])

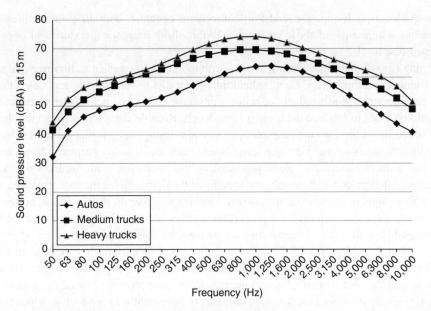

Figure 6.3 Typical roadway traffic noise spectra at 80 km/hour speeds on average pavements (adapted from data in the Traffic Noise Model manual [1])

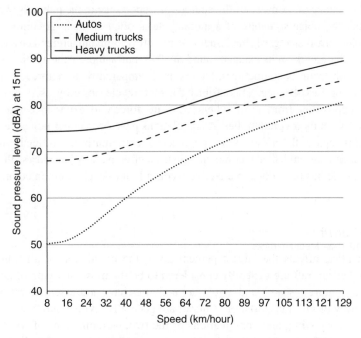

Figure 6.4 Typical roadway traffic noise levels with speed on average pavements (adapted from data in the Traffic Noise Model manual [1])

open-graded asphalt concrete, and Portland cement concrete, with different textures and porosities within each of those categories. Within those categories, texture and porosity dictate noise levels.

Tining (grooves cut into pavements) is a common design practice to improve drainage and friction, and thus safety. Longitudinal tining (parallel to vehicle movement directions) can provide some noise reduction, while transverse tining (perpendicular to the travel directions) tends to produce the highest noise levels. Rumble strips along and in the shoulders of roadways are extreme examples of transverse tining, designs that are used to intentionally increase noise and vibration levels inside vehicles for safety purposes. Noise levels and the frequency content of noise produced by tire interaction with rumble strips are a function of the depth, width, length, and spacing of the tining. Given the appropriate rumble strip dimensions, it is possible for passing vehicles to generate short musical pieces for drivers and the communities, but typical rumble strips generate dominant noise in the 800 Hz range [4], with a goal of generating an increase of noise inside the vehicle of 10–15 dBA, although an increase of 6–12 dBA is usually appropriate for the intended reaction [5].

Tire–pavement noise has become a priority for roadway noise studies in the US and Europe over the past 10 years, but varying pavement types have not proved to consistently provide significant noise reductions, mainly due to pavement wear with time, which minimizes or eliminates the noise reduction benefits of the new surface. This is especially the case for porous or longitudinally tined surfaces that fill with dirt and debris over time to eliminate the porosity or unevenness of the surface that was responsible for the acoustical benefits.

Another category of roadway traffic sources of importance to the public is motorcycles. Experiencing the noise signature of a motorcycle is often an integral component of the drive for many. In that regard, the American motorcycle manufacturer Harley-Davidson applied for a trademark on its exhaust sound in 1994, but withdrew the application in 2000 after opposition from its competitors. Others have tampered with motorcycle mufflers to enhance the aural experience. This affinity for motorcycle noise by users has generated significant opposition from most others who are forced to experience that sound as unwanted noise in their communities. As regulations have been either nonexistent or ineffective in this regard, the noise control engineering community has officially recognized this problem and has published recommendations on effective ways to address the issue [6], but, to date, little has been done to effectively control motorcycle noise in communities.

6.2.2 Aircraft

As aircraft noise affects the largest percentage of the population of all transportation sources, and as aircraft are generally considered to be the most annoying of all transportation sources, more research and regulatory activities have addressed aircraft noise than that of roadway or rail transportation. Aircraft are generally divided into three categories – fixed wing, rotary wing, and combinations of the two. Combinations of fixed-wing and rotary-wing aircraft are part of the new generation of tilt rotor aircraft, which use rotary wings for precise take-off and landing and convert to fixed wings for cruising operations.

The primary noise sources in each case are propulsion systems (engines) and aerodynamic phenomena related to high-speed airflows. Characteristics of fixed and rotary wing sources are described in the following sections.

Fixed Wing

Fixed-wing propulsion systems are either jet- or propeller-based, with most contemporary models being turbojet. Advances in jet design technology since the 1970s, combined with increasingly reduced noise limits by aircraft agencies, most notably the Federal Aviation Administration (FAA) in the US and the International Civil Aviation Organization (ICAO) elsewhere, have resulted in noise reductions in the order of 20 dB in commercial aircraft emissions since these rules have taken effect [7]. The FAA has published noise standards in its Federal Aviation Regulations Part 36, in Title 14 of the Code of Federal Regulations (CFR), and the ICAO has published similar noise standards in its Annex 16, Volume 1. The FAA categorizes aircraft noise level by "stage" numbers and the ICAO categorizes these levels by "chapter" numbers, each beginning with the loudest Stage 1 and Chapter 1. Stage 4 and Chapter 4 are the most restrictive and the most recently introduced. ICAO is proposing a new Chapter 14 standard to begin implementation in 2017, with noise limits 7 dB less than those for Chapter 4 aircraft [8].

Figure 6.5 shows a typical spectrum shape for contemporary aircraft jet noise in the take-off stage behind the aircraft, with maximum levels between 31.5 and 63 Hz, in the range of frequencies that cause rattling of common residential building components [9]. As the jets dominate the noise signature behind the aircraft, higher-frequency engine compressors dominate the noise signature in front of the aircraft. Engine compressors are tonal, dominated by a tone at the blade passage frequency (BPF) of the compressor fan (the rotational

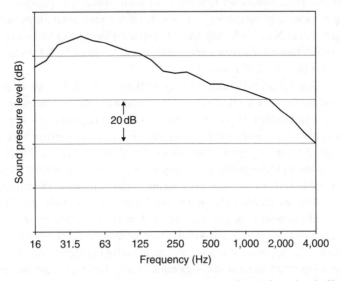

Figure 6.5   Typical fixed-wing aircraft noise spectrum shape (adapted from data in Sharp et al. [10])

fan tip speed multiplied by the number of fan blades). An additional component to the noise signature from air flow around landing gear and irregular surfaces tends to be audible near airports, but is usually not the source of most concerns.

Rotary Wing

Noise signatures from rotary-wing aircraft (helicopters) tend to be dominated by a combination of engine and rotor noise. As rotors provide propulsion and helicopters travel at much slower speeds than most fixed-wing aircraft, engine noise is typically not an issue with communities. Rotor noise is by far the most annoying component of helicopter noise. The impulsive banging or slapping noise typically associated with helicopters is caused by blade–vortex interaction, for which one rotor blade slices through the air and generates vortices in its wake for the next blade to encounter as the rotor spins through the air currents. The turbulent vortex shed by one blade causes a sharp pressure drop, which, in turn, generates the impulsive characteristic rotor noises in a cone in front of and below the plane of the rotors. As the rotor tip speeds can approach the speed of sound, shock waves can be generated, resulting in a sharp, cracking noise that is often annoying to anyone nearby. This type of noise signature occurs mostly while the aircraft is ascending or descending, because that condition sets up the largest pressure differentials between successive rotor blade passages. Although rotor noise levels increase with increasing rotational speed, there is a practical limit to reducing rotor speed to reduce the noise level as the rotors must maintain a minimum speed to provide the required lift force to support the helicopter.

By their design, helicopters fly closer to the ground and, most strikingly different from fixed-wing aircraft, can hover in place for extended periods. Where an airplane flyover event is always relatively short, a helicopter event can be almost indefinitely long. For these reasons, there has been significant movement to limit helicopter operations over heavily populated areas, especially for tourist purposes. A 1999 report about the negative effects of helicopter noise in the New York City area [11] spurred the city to ban tourist helicopters in 2000, and significant efforts have been made by the National Park Service in the US to limit tourist helicopter travel in national parks.

Aircraft noise has historically been rated using the A-weighted L_{dn} descriptor (the day–night equivalent level), and the relevance of L_{dn} on human response continues to be debated for fixed-wing aircraft ratings. However, as helicopters are in the aircraft category, they have also historically been rated using the L_{dn} descriptor. In that light, surveys have shown that a 9-hour (8:00 am to 5:00 pm) L_{eq} in dBA correlates well with annoyance [12]. This is independent of impulsive or non-impulsive events according to these studies. However, another aspect of helicopter noise that adds to its annoyance rating is the generation of rattling of building components. As this results from elevated sound levels below 125 Hz, the dBA scale does not effectively address these issues. A study of small military helicopters showed that noise-induced vibrations and rattles occurred within 150 m of helicopter operations [13]. This study showed that up to a 20 dB drop-off in level was required for helicopter noise to be considered equally as annoying when household components were vibrating or rattling compared with conditions when they were not. This was independent of dBA or dBC ratings.

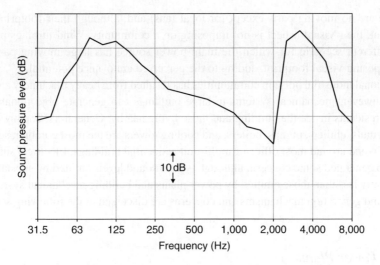

Figure 6.6 Typical rail noise spectrum shape (adapted from data in Hanson et al. [14])

6.2.3 Rail

Rail noise is typically rated as the least annoying of transportation noise sources. The principal sources associated with rail noise are engines or locomotives, wheel–rail interaction, and, for high-speed operations, aerodynamic flow. Figure 6.6 shows a typical transit rail noise spectrum shape, with significant low-frequency energy between 63 and 250 Hz from engine noise and wheel–rail interaction rolling noise, and a peak between 3,000 and 4,000 Hz from squeal produced by cars moving along curved track with radii less than 213 m [15]. For straight track, wheel–rail interaction noise is highly affected by the surfaces of the wheels and rails. Uneven wheel and rail surfaces can add up to 10 dB to the noise signature of a passing train.

The noise source most associated with rail noise annoyance is the warning horn, which generates noise levels in the range of 30 dBA higher than the dominant wheel–rail interaction noise associated with rail cars and 20 dBA higher than the dominant engine noise associated with locomotives [14]. The combination of the high noise levels associated with horns and their rapid onset rate can make them startling in addition to being annoying. Startle begins with onset rates of 15 dB/s, which can occur for high-speed trains as well as horns. High-speed trains – those traveling faster than 150 km/hour – add aerodynamic noise as a dominant source due to the high amount of air turbulence generated by trains traveling at such high speeds. Aerodynamic noise dominates the train's noise signature at speeds higher than 300 km/hour [16].

6.3 Industry and Utilities

Industrial plants have been scattered throughout populated areas for more than a century, but recent renewable energy technologies have created a new generation of environmental noise sources, most notably solar parks and wind turbine farms. Noise complaints from solar parks are rare, because the sound generated by them is typically at a low level (as there are no moving parts except for local fans) and, although their footprint can be significant, their visual aspect is not imposing on a community. Wind turbines provide a much different scenario for which the main power source has large moving parts and a more imposing visual footprint, adding to the perceived annoyance potential.

Traditional industrial noise in communities has resulted from heavy machinery and power plants; however, mechanical systems in large buildings can generate objectionable noise both for residents inside the building and those living nearby. Generators, compressors, air handling units, chillers, pumps, boilers, and cooling towers are the most common sources of concern in and around most office or multi-unit residential buildings. Electrical substations have also generated some concern, although typical sound levels emitted by substations are significantly less than those emitted by power plants and building mechanical systems. The basic sound generation mechanisms and concerns are discussed in the following sections.

6.3.1   Power Plants

Much has been studied and written with regard to noise generated by mechanical and electrical equipment in fossil fuel power plants since the 1970s. Most heavy equipment manufacturers have noise data included in their specifications, which can be used in models to predict whether the noise generated by this equipment will violate municipal limits or generate environmental impacts that need to be addressed by noise control designs. The common noise sources among this equipment can be categorized by mechanical, fluid flow, and electromagnetic origins.

Dominant mechanical noise sources are typically rotating components of machinery, such as fans or gears, which generate tones directly related to their rotational speeds. Whereas gears generate noise from direct contact with solid materials, fans generate noise from their induced fluid flows. Interactions of these fluid flows with system components can amplify the sounds and generate new sounds from turbulence and flow restrictions. As an example, turbine engines have sets of rotating fans (rotors) in close proximity to sets of stationary blades (stators), designed to smooth the flow for greater efficiency. The interaction of fluid flow with this system of rotating and stationary blades (known as rotor–stator interaction) generates dominant tones at specific frequencies associated with the blade passage frequency and its harmonics, with the strength of individual harmonics depending on the number of rotating and stationary blades in the system. Monitoring the acoustic characteristics (spectral shapes) of these tones is key to maintenance evaluations and early detection of issues that can avoid catastrophic equipment failure.

Fluids (in both liquid and vapor forms) flowing through ducts and pipes at high rates can generate significant noise levels, mainly as a result of the turbulence of the flows through those conduits. Sharp bends or changes in cross-sectional area in a flow path exacerbate noise and lower efficiency, and so designing systems to minimize flow turbulence also minimizes noise generation in those systems.

Dominant electromagnetic noise sources are associated with electric motors, generators, lighting components, and transformers. Electric motors and generators typically have strong magnetic forces produced by the mechanical and electromagnetic properties of system components, most notably rotating equipment that generates fluid flows. Fluorescent lights and transformers function through the property of magnetostriction, in which the dimensions of components change when their materials are magnetized. The rate at which magnetostriction occurs is related to the electrical current flowing through the system. As most electrical systems are based on constant 50 and 60 Hz currents, vibrations are generated in components of these systems that are proportional to the base alternating current frequency, producing the characteristic humming sounds associated with these types of sources (at the fundamental frequency of 50 or 60 Hz and harmonics of those frequencies).

All of these noise sources can also generate vibrations in machinery components, which, when these components are rigidly attached to building elements, can be transmitted through a building's structure to remote locations. In addition, resonances of equipment and/or attached building elements can amplify this vibrational energy, so all systems should be designed with this in mind to avoid the potential problems associated with this. Vibrating materials can generate their own noise, which can be introduced at locations far from the source through a complicated chain of rigid connections and resonances. The discussion of noise control in Chapter 8 covers the most practical ways to avoid these types of issues.

6.3.2   Wind Farms

Wind turbines have been used throughout recorded history to generate mechanical energy, and for more than a century to generate electrical energy. Environmental and political issues in recent history have accelerated the proliferation of wind turbine electrical energy production throughout the world. Because of this, wind turbine farms have been, and continue to be, developed in rural areas where background noise levels are typically low, thus creating the issue of wind turbine noise being introduced into quiet residential areas. This has become more of a problem since the 1990s because wind turbine sizes have dramatically increased to roughly 100 m hub heights (at the center of the turbine blade and generator assemblies) with turbine blade diameters in the 100 m range, introducing not only an acoustical but also a visual aspect. The principal types of complaints about wind farms are related to noise and sunlight flickering through the rotating blades, with flicker occurring in limited time-frames depending on the location and season.

There are two categories of noise sources associated with wind turbine generators – mechanical and aerodynamic. Mechanical sources are in the generator compartment behind the blades, consisting of a gearbox, a generator, and power train controlling mechanisms. Unless there are maintenance issues, mechanical noise associated with wind turbines is usually not audible beyond 100 m, although the tones associated with the gearbox are usually the dominant noise sources from the generator compartment [17]. In contrast to most other environmental noise sources, aerodynamic noise far outweighs all other sources for wind turbines, with the potential for them to be audible more than 1 km away. Aerodynamic noise associated with wind turbines is caused by the large blades cutting through the air while being exposed to the turbulence associated with the local winds and the vortices shedded by subsequent blade passages. The principle is similar to that associated with helicopter rotor noise but at much slower rotational speeds.

The aerodynamic noise signature associated with an operating wind turbine has two characteristics – a mid-frequency (500–1,000 Hz) swishing sound from the turbulence of each blade cutting through the air and an amplitude-modulated level associated with the rotational speed of the blades. Many wind turbines installed before the 1990s had two blades that were oriented downwind of their support towers. This downwind orientation introduced a wake in the air flow pattern encountering the rotating blades, causing a dominant low-frequency (<100 Hz) thumping component that could be annoying to listeners downwind of the units [18,19]. Since that time, most wind turbines have been designed with the rotors upwind of their supporting towers, thus eliminating that aspect of their acoustic signatures. Measurements and calculation models have shown that turbines having upwind rotors can be 10–20 dB quieter at all frequencies than those with downwind rotors [20]. Figure 6.7 shows the modeled effect of the distance between the tower and the blade plane for noise generation in both upwind and downwind designs. This assumes a three-blade 5 megawatt (MW) turbine with a hub height of 100 m and a blade diameter of 127 m. Besides the significant difference in sound levels generated between upwind and downwind orientations, increasing the distance between the blade plane and the tower also significantly reduces this type of noise generation at frequencies > 20 Hz.

Wind turbines begin to operate at minimum wind speeds of around 4 m/s, known as the cut-in speed. Their electric power and noise output increase with increasing wind speeds up to a point above which the generated power and noise levels remain fairly constant, typically around 12 m/s [22]. Most contemporary 1.5 MW turbines are rated in the 100 to 105 sound power level range.

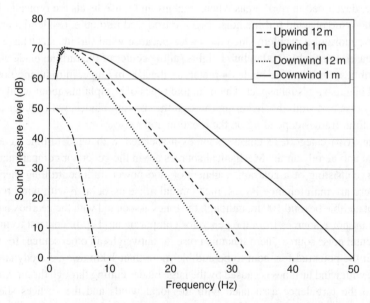

Figure 6.7   Effect of distance and location of turbine with respect to the support tower at 400 m downstream (adapted from data in Madsen [21], with permission of Multi-Science Publishing Co. Ltd.)

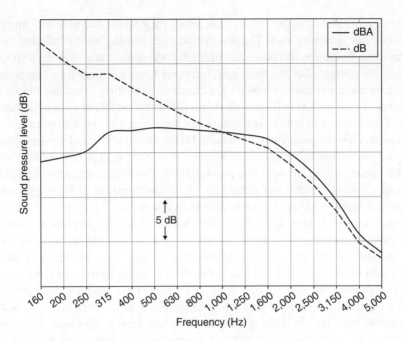

Figure 6.8   Typical wind turbine noise spectrum shape at 500 m (adapted from data in Bowdler and Leventhall [23], with permission of Multi-Science Publishing Co. Ltd.)

Figure 6.8 shows a typical wind turbine noise spectrum shape at a horizontal distance of 500 m, but bear in mind that the source is typically 100 m or more above that elevation, so atmospheric effects play a major role in sound propagation between a wind turbine generator and a listener close to the ground. The figure also shows the unweighted and A-weighted sound pressure levels to show the low-frequency contribution that is discounted by the human hearing mechanism but not by building structures. Perceptible floor vibrations in typical house constructions can be generated by sound pressure levels above 80 dB at 10 Hz and above 100 dB at 100 Hz [24]. Perceptible wall and ceiling vibrations in typical residential structures can occur above 75 dB at 16 and 31.5 Hz, and above 80 dB at 63 Hz [9]. According to the data in Figure 3.5, sound energy at these levels can also generate feelings of vibrations in the head for frequencies > 31.5 Hz. However, measurements of noise from wind farms consistently show sound pressure levels at least 10 dB below this threshold at 300 m [25–28].

Wind farm developers have encountered strong public opposition to their projects since the 1990s because of the plethora of published articles and anecdotes addressing the potentially negative health effects related to wind turbine noise exposure. Many research studies have been performed to address these issues with varying agendas, but there are consistencies in the scientific study conclusions. The consistent conclusions are that noise from wind turbines has significant contributions in the mid-frequency range (500–1,000 Hz), with modulations in the 1–3 Hz range from blade rotation through unstable turbulent air flows at the different elevations.

It is generally agreed that this noise can be annoying at levels above the background, but that is where the agreement ends. The disagreement lies with the interpretation of the infrasound and low-frequency characteristics of the turbine noise rather than the audible midfrequency components. For the most part, infrasound and low-frequency components of wind turbine noise signatures are below the threshold of hearing, especially below 50 Hz, at distances exceeding 300 m from the turbines [25–28]. One school of thought states that no significant infrasound or low-frequency noise is generated by wind turbines and, as any infrasound and low-frequency noise below 50 Hz associated with wind turbines more than 300 m away are below the threshold of hearing, that criterion alone eliminates the possibility of the noise causing any issues for people [29,30]. Other arguments admit that, although infrasound associated with wind turbines is below the threshold of hearing, the human body can still sense these signals and be affected by them [31].

As discussed in Chapter 3, cochlear inner hair cells are the carriers of most sound energy information to the brain that is interpreted as audible sound and infrasonic signals are minimized through that channel. However, outer hair cells are up to 40 dB more sensitive than inner hair cells to signals below 20 Hz (due to their direct contact with the shearing motions of the tectorial membrane in response to infrasound) and their stimulation can produce sensations of fullness or pressure in the ear as well as tinnitus [32]. Although this argument may explain some symptoms that some people experience who live near wind farms, it does not explain why other common sources of infrasound, both natural and artificial, that are of a similar level to that produced by wind farms do not affect people in the same way. This limits the validity of this argument, as most other sources of regularly occurring environmental infrasound are of similar, if not higher, levels to those produced by wind farms [33].

Beyond this, there is also disagreement regarding whether the periodic nature of wind turbine noise (due to the rotating blades) can be classified as infrasound at all or whether it is a modulation of sound with its primary components in the mid-frequency range. Amplitude modulation is a periodic pulsing variation of sound level with time. This tends to be more pronounced at night when wind currents are more stable and stronger at higher elevations than they are closer to the ground. Because modern wind turbines are tall enough to encounter these changes, wind speeds at elevations of 150 m (the typical maximum blade tip elevation for a 1.5 MW turbine) can be much higher than wind speeds at elevations of 50 m (the minimum blade tip elevation). Turbine blades rotating through these variations in wind currents change their angles with respect to the wind currents (known as the angle of attack), potentially setting up stall conditions in the blades (for which turbulent air layers separate from the blades causing them to lose their lift and power-generating capabilities), thus amplifying the modulations from roughly 1 dB during daytime hours to more than 5 dB at night [34]. This kind of amplitude-modulated sound can be more annoying than unmodulated signals at the same sound level, which is one theory as to why wind turbine noise may be more annoying than other environmental sounds (such as transportation or industrial sources), especially at night [35]. Although modulation depth (the range of variation of the modulated signal) has been identified as a factor in wind turbine noise annoyance, the annoyance rating of modulated sounds has been found not to increase for modulation depths > 3 dB [36].

Studies on public perception of wind turbine noise have primarily used questionnaire-based surveys, and these have yielded some common trends regarding the non-acoustic aspects of wind farms. These trends are based on economic, legal, visual, and attitudinal aspects. Also, study results can be biased according to the people chosen to complete the surveys as well as those who choose to reply. That being the case, there appears to be wide agreement that those who economically benefit from wind turbines are less likely to complain about them and those involved in legal actions against a wind turbine development are more likely to complain. Annoyance has also been correlated with the visual aspect of wind turbines and strong negative attitudes toward the turbines [37–40]. Of the dose–response relationships derived for wind turbine noise annoyance, Figures 5.1 and 5.2 in Chapter 5 show that wind turbines are generally rated to be more annoying at lower sound levels than transportation sources [41].

The most controversial aspect of the effects of wind turbine noise is health. A host of stress-related symptoms have been casually linked with wind turbine noise, with mixed acceptance from the scientific community. There have also been studies on sleep disturbance from wind turbine noise [42]. These studies reference the impulsive nature of wind turbine noise, but, as for infrasound, there is no consensus that wind turbines generate impulsive signals and this impulsive characterization of wind turbine noise is thought by some to be an interpretation of amplitude modulation.

The most extreme health claims related to wind turbine noise stem from a book written by an American pediatrician which introduced "wind turbine syndrome," allegedly caused by an illness called visceral vibratory vestibular disturbance (VVVD). This illness is defined as "a sensation of internal quivering, vibration, or pulsation accompanied by agitation, anxiety, alarm, irritability, rapid heartbeat, nausea, and sleep disturbance [43]." Other symptoms mentioned include headache, dizziness, tinnitus, ear pressure, vertigo, visual blurring, concentration and memory problems, and panic episodes. Although some of these symptoms may be supported by claims mentioned earlier regarding the potential effects associated with outer hair cells being capable of sensing low-frequency signals that are not processed as audible sound in the brain, the majority of VVVD symptoms have been associated with many other illnesses, particularly those induced by stress.

The principal criticism of wind turbine syndrome by the scientific community is that the conclusions of the study are based on telephone interviews with 10 families from different geographic areas and at different distances (between 305 and 1,500 m) from different wind farms, chosen only because they exhibit the most extreme symptoms. The ages of the participants vary from less than 1 to 75 years. Independent of other claims that can be made about the validity of the hypothesized methods by which VVVD is realized, the only consistency in this study is the condition that all of the participants have extreme symptoms (which is the reason they were chosen for the study). Although VVVD is a medical diagnosis, the author of the study neither met nor examined any of the participants. There were no control groups or accommodation for confounding factors in the study. In most ways, the methods used to present the case for this malady are unscientific.

Besides the lack of a scientific basis to this study, there is support for the notion that many people claiming to be affected by VVVD may be experiencing what is called a nocebo effect, for which an anticipated negative effect is realized after being exposed to information about the potential negative effects associated with a specific agent [44]. This is contrasted from the better-known placebo effect, for which the suggestion of a positive effect results in that outcome.

There is general agreement in the scientific community that wind turbine noise can be annoying to people, especially within 300 m of a turbine, but, as outlined previously, there are many facets to that annoyance for which there is significant disagreement, so more defensible research is needed to clarify these issues. In addition to defining the scientific bases for any potential effects, there also needs to be a consensus on how wind turbine noise is rated. The basis for this is in not only defining acceptable levels but also relevant descriptors. Some research has shown that A-weighted sound levels adequately describe human reactions to wind turbine noise [45], while others say that as low-frequency issues are the prime concern for wind turbine noise, they are not properly addressed by A-weighted levels and that C-weighted or unweighted levels need to be included [28,46]. Key issues that need to be defined and agreed upon relate to how wind turbine noise is rated and whether those rating methods need to be different from rating methods used to assess potential impacts from other noise sources.

6.3.3   Electrical Power Systems

For public utility alternating current electrical power systems, there are two principal sources of noise that have been potential annoyance issues for people – corona and transformers. Corona noise is associated with ionization of fluids surrounding electrically charged conductors in transmission lines, and is usually only noticeable in damp conditions as a cracking or hissing noise, in addition to a tonal low-frequency hum, which is also characteristic of transformer noise. The hum is usually dominant at twice the power line frequency (100 Hz for a 50 Hz system and 120 Hz for a 60 Hz system). There are also typically lower-level tones at the power line frequency itself and at higher harmonics. In the US, transformers are required to meet the audible noise limits prescribed in the National Electrical Manufacturers Association (NEMA) TR 1 standard [47]. Noise levels to be guaranteed by manufacturers are listed by transformer power ratings. Calculations to determine noise levels associated with these systems can be found in the Electrical Power Research Institute (EPRI) *AC Transmission Line Reference Book* [48].

6.4   Personal/Recreational Sources

Environmental sound sources are pervasive in most societies around the world, and we often have little, if any, control over our exposures to those sources, especially when those sources are associated with occupations. The types of sources we have the most control over are those we are exposed to by choice, through our recreational activities. As was mentioned in Chapter 5, one of the key elements of stress related to noise exposures is whether or not the listener has control over the source. In the case of recreational sources, listeners have the ultimate control, as they are choosing this exposure. Therefore, annoyance or stress-related issues should not be associated with recreational noise exposures.

The only exception to that rule is the case of the listener being exposed to someone else's chosen noise, for which noise ordinances have been created.

From the perspective of the listener who chooses his or her exposure, the main concern is noise-induced hearing loss because many recreational sources generate sound levels that exceed those thresholds. The most common of these sources are firearms, public performances, personal listening devices, toys, and power tools. None of these issues is new, with studies proving the potential hearing hazards of these sources for more than the past 50 years.

6.4.1 Firearms

Of the many recreational sources people exposed themselves to, firearms have the greatest potential to cause noise-induced hearing loss because of the high sound pressure levels associated with them (up to 172 dBA at the ear of the listener [49,50] for rifles and shotguns) combined with their impulsive nature. As was mentioned in Chapter 3, the acoustic reflex in the middle ear is not activated quickly enough to provide protection against impulsive sources. Therefore, the entire signal passes through the hearing mechanism at a level that can clearly damage cochlear hair cells, with the possibility of causing structural damage to the eardrum and middle ear ossicles (known as acoustic trauma). As one study showed more than 30 years ago, isolated exposure to recreational gunfire caused changes in auditory thresholds similar to those caused by 20 years of daily occupational exposures averaging 89 dBA [51]. An exposure to a single shot from a high-powered rifle (with a maximum level of 165 dB) is equivalent to a 90 dBA average 8 hour/day exposure level for 1 week, and 50 shots is equivalent to that same exposure level for 1 year [52]. Self-rated hearing handicaps were also reported as being significantly higher for males, older individuals, and blue collar workers, most likely due to their use of more powerful firearms and their less frequent use of hearing protection compared with females and white collar workers [53].

6.4.2 Public Performances

Common public performances that generate the highest noise levels are musical concerts and race car tracks. It is accepted from many years of studies that sound levels in nightclubs routinely exceed 100 dBA and sound levels during live concerts routinely exceed 120 dBA [49]. Before the 1990s, most concert sound systems consisted of stacks of loudspeakers on either side of a stage, resulting in the highest exposure levels for those closest to the stage. Since that time, most concert systems have consisted of arrays of loudspeakers facing in different directions, each capable of being adjusted to spread sound at even levels throughout the audience, and so concert-goers sitting farthest from the stage are exposed to the same sound levels as those close to the stage (and sometimes higher). Peak sound levels at these concerts can exceed 140 dB, certainly high enough to cause temporary hearing threshold shifts, but unless one attends many concerts they are usually not concerns for permanent threshold shifts.

Spectator sports typically don't expose attendees to potentially damaging sound levels, but crowd noise at large events has been measured to exceed 110 dBA [54]. One exception to this rule is auto racing, for which spectators are exposed to peak sound levels exceeding 120 dBA and average (L_{eq}) levels exceeding 100 dBA from passing cars over several hours [55,56].

6.4.3   Toys/personal Listening Devices

Noise sources that we have the greatest control over are toys and personal listening devices. Even after more than 50 years of studies relating high levels of noise exposure to hearing loss, toys continue to be manufactured that generate potentially hazardous levels. The most extreme of these are toy cap guns and firecrackers, measured to produce impulsive noise levels in excess of 150 dB at 50 cm for cap guns and 170 dB at 3 m for firecrackers [52,57]. Studies have revealed that more than 10% of children in the US alone have noise-induced threshold shifts in one or both ears, and toys introduce a significant contribution to that statistic [58]. The consumer product safety communities are now providing limits on sound levels produced by toys in Europe (through Standard EN 71-1 [59]), Canada (through the Hazardous Product Act), and the US (through ASTM F963 [60]), as well as at the international level with the International Organization for Standardization (through ISO 8124-1 [61]); however, these limits are often exceeded and are only based on unenforceable standards [62]. The limits in these standards vary widely, often using several criteria such as A-weighted averages and C-weighted or unweighted peak levels. The European standard also separates categories of toys and has different limits based on exposure durations, with three duration categories of less than 5 s, between 5 and 30 seconds, and more than 30 s. Average limits are generally in the 80–90 dBA range and peak limits are in the 110–125 dBC range for impulsive sources (with the higher limits for explosive toys such as caps or guns), mostly at a distance of either 25 or 50 cm. As the requirements in these standards have changed significantly over the past 10 years, it is best to reference their most current versions.

Another aspect of the noise-induced hearing loss potential of toys is that small children have different hearing sensitivities than adults. The ear canal resonance mentioned in Chapters 2 and 3 is relevant to an adult's ear canal. As this resonance is based on the size of the canal and children's ear canals are smaller than those of adults, this resonance is shifted to a higher frequency range for children younger than 7 years of age [63]. In fact, the ear canal resonance frequency at birth is roughly twice that of an adult, at nearly 6,000 Hz [64]. This has implications for noise-induced hearing loss vulnerability differences between infants and adolescents, especially considering the different types of toys catered to these age groups.

Arguably the most controllable recreational noise source is the personal listening device. Many studies have been performed since personal music systems were developed in the 1970s. Early versions of these devices generated sound levels up to 128 dBA at maximum settings and up to 111 dBA at moderate settings [65]. Although current earbuds and headphones generally produce slightly lower levels (a 2007 study of nine devices and 20 types of earphones yielded maximum playback levels of 101–107 dBA using standard measurement techniques, but other factors could increase the maximum levels to 125 dBA [66]), they are still capable of producing sound levels that can cause sensorineural hearing loss from prolonged exposures. One researcher has estimated that a 15-year old who uses a personal music player at 90% of maximum volume every day for 2–4 hours for 10 years would have accelerated the aging of his hearing mechanism by 30 years when compared with standards from presbycusis curves [67]. There are mixed results in terms of studies attempting to quantify noise level exposures and associated hearing loss risks from personal music players, but males in general tend to use them for longer time periods and at levels on average 5 dBA higher than females [68]. Although these studies show the potential for noise-induced hearing loss from sound exposures associated with these devices alone, other noise exposures may also contribute to their hearing loss and these factors are often not considered in these studies [69]. Some of these confounding factors (such as history of other noise exposures) are eliminated, though, as the studies mostly use young adults (under the age of 30).

A few more points are worth noting. First, there is a broad range of output sound levels associated with personal music systems, with measured variations of up to 20 dBA for the same loudness settings [70]. Second, the background sound level will dictate the maximum level played through the systems. Higher background levels tend to encourage higher system level settings. Lastly, since many people use personal music systems when exercising, it is worth noting that exercise can make one more vulnerable to hearing loss when exposed to high noise levels, due to reduced blood flow to the cochlea during exertion [71]. These susceptibilities are further increased, considering that most workout facilities have high background noise levels which encourage personal music system users to raise their volumes.

As discussed in Chapter 8, hazardous conditions for developing noise-induced hearing loss begin with long-term exposures averaging 85 dBA. With that in mind, it has been found that personal listening devices tend to be set to levels exceeding 85 dBA when background levels exceed 72 dBA [72]. This provides a guideline limit for background sound levels in environments in which personal listening devices are typically used.

6.4.4 Appliances/Tools

Typical household appliances don't emit sound levels near the potentially hazardous range. Power tools that may approach these levels are woodworking tools, chain saws, and leaf blowers. Leaf blowers have been measured up to 112 dBA at the user and chain saws have been measured up to 116 dBA at the operator's ear [49]. One study showed that people involved in woodworking activities were 31% more likely to have significant hearing loss than those who have not done woodworking, with an additional 6% increased risk with every 5 years of exposure [73]. In this case, hearing loss was defined as an average of more than a 25 dB threshold shift between the frequencies of 500 and 4,000 Hz in either ear.

6.5 Hums

The topic of hums was introduced in Chapter 3 as phantom sources. Hums are typically defined as sounds perceived by a small percentage of the population, with no obvious source. They have gained attention over the past 40 years due to their mysterious nature; however, most hum cases have been solved. Hums can be divided into two categories – acoustical and non-acoustical. All hum cases that have been solved have been purely acoustical in nature, for which a microphone connected to a sound analyzer can clearly detect the signal, which then can be traced to a remote source having the same spectral characteristics. Non-acoustical hums cannot be measured with sound analyzers and they therefore provide the greatest challenges for source identification. In any case, the identification of the source is crucial to eliminating the problem.

The sources of acoustical hums are often industrial [74] but have been known to be animals or fish. An example is the hum in Sausalito, California, in the 1980s that was attributed to the mating calls of the toadfish [75]. Some hum cases have been attributed to infrasonic tones [76,77] and others have been attributed to tinnitus [78,79], but most do not fall into the tinnitus category because the perceived sound is often localized geographically while tinnitus (not resulting from an external stimulus) is independent of location.

Although acoustical hum sources are much easier to identify than non-acoustical sources, there are still many cases for which acoustical sources cannot be identified. In many of those cases, however, the source can be localized to a particular facility, but it cannot be specifically identified because the facility is inaccessible. One recent example of this is in Calgary, Alberta. In this case, a 43 Hz acoustic signal was measured, but not consistently enough to identify the source [80]. Another case currently under investigation is the Windsor Hum in Windsor, Ontario, for which a 35 Hz tone characteristic of a blast furnace has been measured and the source is thought to be located in an inaccessible industrial facility on a nearby island [81,82].

Non-acoustical hums have received the most publicity because they support the unsolved mystery industry. As the sources of acoustical hums can often be easily identified, the sources of non-acoustical hums are most often not. One can be sitting next to someone experiencing a hum without hearing it or sensing it on instruments, an aspect that has caused many of those who experience hums to be publicly ridiculed. Medical issues accompanying hums have been reported in many cases, including headache, nausea, diarrhea, fatigue, and memory loss. Some also report feeling vibrations in addition to hearing the sound. In most cases, the sounds suddenly become noticeable and, although they vary in description, they are most often characterized as sounding like an idling diesel engine with a pulsed intensity.

Hums have been reported in all parts of the world over the past 40 years, most notably in Europe, Australia, and the US. Two large hum studies have been commissioned by governmental funds in the US over the past 20 years – one in Taos, New Mexico, in 1993 and another in Kokomo, Indiana, in 2003. Extensive acoustical, electromagnetic, and seismic measurements were taken by a team of researchers in Taos without any conclusions. The only abnormality reported by the team from the University of New Mexico was that stray electromagnetic fields were strong at distances away from the power lines at odd harmonics from 180 to 2,000 Hz, but no acoustic or seismic signals stood out [83] and the source of the hum has not been identified.

Acoustic measurements in Kokomo revealed dominant tones at 10 Hz (with strong tones at higher harmonics as well) and 36 Hz that could be attributed to specific pieces of equipment in two industrial facilities [84]. Noise control measures were instituted for each of those pieces of equipment and the dominant tones were eliminated; however, most residents continued hearing the hum after the equipment had been quieted. Many affected residents mentioned in pre-monitoring interviews that they had also experienced electrical issues in their homes (such as appliances suddenly failing, light bulbs bursting, and cars having remote starters unexpectedly starting in garages) since they started hearing the hum. Therefore, electromagnetic field readings were taken in their homes in addition to the acoustical readings. In most of these cases, electromagnetic field strengths were elevated in and around those homes. In addition to the electrical problems reported, there were also dead trees and bushes near these houses, and it was reported that pets and birds were reacting to the hum with erratic behaviors at the same times people were sensing the signals. The focus of the study had been the sound, so the identification of acoustic sources satisfied those commissioning the study and no further investigations were performed. It should be noted, however, that a former steel plant site in the town had been formally identified by the US Environmental Protection Agency as having contaminated groundwater and soils, and local residents interviewed for the study began hearing the hum and experiencing other symptoms around the time the clean-up process began.

Theories abound regarding the source of non-acoustic hums [85], from military communications systems to extra-terrestrials, but none of them has been verified. Although evidence is leaning toward local electromagnetic issues, credible research needs to be performed to verify that. The seemingly unsolvable nature of these cases adds to the frustration of those experiencing the unexplainable symptoms, which can certainly exacerbate the situation.

6.6 Acoustic Weapons

Acoustic weapons have been the subject of folklore throughout recorded history. From the biblical tales of the sounds of ram's horns causing the walls of Jericho to collapse to the tales of Gavreau's infrasonic devices causing men to become invalids (mentioned in Chapter 4), sound has been glorified as a potentially dangerous tool. The subject of acoustic weapons would not be complete without mentioning the mythical brown note, a pure tone that, when reproduced anywhere near a person would allegedly cause involuntary fecal incontinence. This is based on the theory that exposure to the resonance frequency of the large intestine would result in the uncontrollable condition. This would be a welcome tool for gastroenterologists to solve the serious problem of severe impaction/constipation if it was a real possibility. Unfortunately, the human large intestine is connected to other tissues having different densities that make a pure resonance situation isolated at the large intestine difficult, if not impossible, to achieve at safe levels.

The premise of non-lethal acoustic weapons has been in active development and use over the past century, since electronic sound reproduction was developed. In *Sonic Warfare*, Steve Goodman explores the range of potentially strong effects that sound can have on people, from eliciting fear to subliminal manipulation [86]. The legitimacy of these claims has yet to be verified but legitimate acoustic weapons do exist, most notably acoustic cannons that consist of arrays of loudspeakers in a single package designed to focus high-intensity sound waves at specific targets. These have been used effectively from dispersing unruly crowds to deterring pirates at sea, generating signals in excess of 120 dBA at 100 m in the frequency range of highest human sensitivity (around 2,500 Hz). They can also be used as effective warning devices at large distances. Their array design facilitates a narrow beam of acoustical energy that cannot be achieved by a single loudspeaker. The negative effects associated with exposure to these devices are limited to annoyance and aural pain as sound levels associated with other severe physiological effects (such as nystagmus, vestibular issues, nausea, blurred vision, and organ damage) are much higher than those that can be generated by these systems.

Sounds to do not have to be loud to manipulate people. The array technology used to develop high-intensity focused sound has also been used in more subtle ways in retail stores to focus aural messages on people as they pass by different items in the store [87]. The most subtle acoustic manipulations are performed through music, for which catchy tunes are used as branding tools through earworms [88]. These are discussed in more detail in Chapter 7.

References

[1] Menge, C.W., et al. *FHWA Traffic Noise Model, Version 1.0, Technical Manual. Report Number FHWA-PD-96-010*. Washington, DC: Federal Highway Administration, 1998.

[2] Rochat, J.L., et al. *FHWA Traffic Noise Model (TNM) Pavement Effects Implementation Study: Progress Report 1. Report Number DOT-VNTSC-FHWA-12-01*. Washington, DC: Federal Highway Administration, 2012.

[3] Donavan, P.R. *Comparative Measurements of Tire/pavement Noise in Europe and the United States – NITE Study. Report Number FHWA/CA/MI-2006/09*. Sacramento, CA: California Department of Transportation, 2006

[4] Sexton, T.V. *Evaluation of Current Centerline Rumble Strip Design(s) to Reduce Roadside Noise and Promote Safety. Report No. WA-RD 835.1*. Olympia, WA: Washington State Department of Transportation, 2014.

[5] Torbic, D.J., et al. *Guidance for the Design and Application of Shoulder and Centreline Rumble Strips. NCHRP Report 641*. Washington, DC: Transportation Research Board, 2009.

[6] Maling, G.C. and Vanchieri, C. *Noisy Motorcycles – An Environmental Quality-of-life Issue*. Institute of Noise Control Engineering of the USA, 2013.

[7] National Academy of Engineering, *Technology for a Quieter America*. Washington, DC: The National Academies Press, 2010.

[8] Transportation Research Board. *Critical issues in Aviation and the Environment 2014. TRB Circular E-C184*. Washington, DC: TRB, April 2014.

[9] Acoustical Society of America. *ANSI S12.2-2008. Criteria for Evaluating Room Noise*. New York: American Institute of Physics, 2008.

[10] Sharp, B.H., Gurovich, Y.A. and Albee, W.W. *Status of Low-Frequency Aircraft Noise Research and Mitigation. Wyle Report WR 01-21*. Arlington, VA: Wyle Acoustics Group, 2001.

[11] Cunningham, C. *Needless Noise: The Negative Impacts of Helicopter Traffic in New York City and the Tri-State Region*. New York: Natural Resources Defense Council, 1999.

[12] Fields, J.M. and Powell, C.A. (1987). "Community reactions to helicopter noise: Results from an experimental study." *Journal of the Acoustical Society of America*, 82(2): 479–492.

[13] Schomer, P.D. and Neathammer, R.D. (1987). "The role of helicopter noise-induced vibration and rattle in human response." *Journal of the Acoustical Society of America*, 81(4): 966–976.

[14] Hanson, C.E., et al. *Transit Noise and Vibration Impact Assessment. Report number FTA-VA-90-1003-06*. Washington, DC: Federal Transit Administration, 2006.

[15] Nelson, J.T. *Wheel/Rail Noise Control Manual. TCRP Report 23*. Washington, DC: Transportation Research Board, 1997.

[16] Hanson, C.E., Ross, J.C., and Towers, D.A. *High-Speed Ground Transportation Noise and Vibration Impact Assessment. Report No. DOT/FRA/ORD-12/15*. Washington, DC: Federal Railroad Administration, 2012.

[17] Jakobsen, J. (1990). "Noise from wind turbine generators. Noise control, propagation, and assessment." *Proceedings of Internoise 90*, Gothenburg, Sweden, 303–308.

[18] Hubbard, H.H., Grosveld. F.W. and Shepherd, K.P. (1983). "Noise characteristics of large wind turbine generators." *Noise Control Engineering Journal*, 21(1): 21–29.

[19] Hubbard, H.H. and Shepherd, K.P. (1991). "Aeroacoustics of large wind turbines." *Journal of the Acoustical Society of America*, 89(6): 2495–2508.

[20] Hubbard, H.H. and Shepherd, K.P. *Wind Turbine Acoustics*. NASA Technical Paper-3057 DOE/NASA/20320-77. Hampton, VA: National Aeronautics and Space Administration, 1990.

[21] Madsen, H.A. (2010). "Low frequency noise from wind turbines mechanisms of generation and its modelling." *Journal of Low Frequency Noise, Vibration and Active Control*, 29(4): 239–251.

[22] van den Berg, G.P. *The Sounds of High Winds: the Effect of Atmospheric Stability on Wind Turbine Sound and Microphone Noise*. PhD thesis, University of Groningen, 2006.

[23] Bowdler, R. and Leventhall, G. *Wind Turbine Noise*. Essex, UK: Multi-Science Publishing Co. Ltd., 2012.

[24] Shepherd, K.P., Grosveld, F.W. and Stephens, D.G. (1983). "Evaluation of human exposure to the noise from large wind turbine generators." *Noise Control Engineering Journal*, 21(1): 30–37.

[25] O'Neal, R.D., Hellweg, R.D. and Lampeter, R.M. (2011). "Low frequency noise and infrasound from wind turbines." *Noise Control Engineering Journal*, 59(2): 135–157.

[26] Jung, S.S., et al. (2008). "Experimental identification of acoustic emission characteristics of large wind turbines with emphasis on infrasound and low frequency noise." *Journal of the Korean Physical Society*, 53(4): 1897–1905.

[27] Hayes Mckenzie Partnership Ltd. *The Measurement of Low Frequency Noise at Three UK Wind Farms*. UK Department of Trade & Industry, 2006.

[28] Kamperman, G.W. and James, R.R. (2008). "Simple guidelines for siting wind turbines to prevent health risks." *Proceedings of Noise-Con 2008*, Dearborn, Michigan.

[29] Leventhall, G. (2006). "Infrasound from Wind Turbines – Fact, Fiction or Deception." *Canadian Acoustics*, 34(2): 29–36.

[30] Jakobsen, J. (2005). "Infrasound emission from wind turbines." *Journal of Low Frequency Noise, Vibration and Active Control*, 24(3): 145–155.

[31] Salt, A.N. and Hullar, T.E. (2010). "Responses of the ear to low frequency sounds, infrasound and wind turbines." *Hearing Research*, 268: 12–21.

[32] Salt, A.N. and Kaltenbach, J.A. (2011). "Infrasound from wind turbines could affect humans." *Bulletin of Science, Technology & Society*, 31(4): 296–302.

[33] Turnbull, C., Turner, J., and Walsh, D. (2012). "Measurement and level of infrasound from wind farms and other sources." *Acoustics Australia*, 40(1): 45–50.

[34] van den Berg, G.P. (2005). "The beat is getting stronger: the effect of atmospheric stability on low frequency modulated sound of wind turbines." *Journal of Low Frequency Noise, Vibration and Active Control*, 24(1): 1–24.

[35] van den Berg, F. (2009). "Why is wind turbine noise noisier than other noise?" *Proceedings of Euronoise 2009*, Edinburgh.

[36] Fiumicelli, D. *Wind Turbine Amplitude Modulation: Research to Improve Understanding as to its Cause and Effect*. London: Temple Group, 2013.

[37] Johansson, M. and Laike, T. (2007). "Intention to respond to local wind turbines: The role of attitudes and visual perception." *Wind Energy*, 10(5): 435–451.

[38] Taylor, J., et al. (2013). "The influence of negative oriented personality traits on the effects of wind turbine noise." *Personality and Individual Differences*, 54(3): 338–343.

[39] Devine-Wright, P. (2005). "Beyond NIMBYism: Towards an integrated framework for understanding public perceptions of wind energy." *Wind Energy*, 8: 125–139.

[40] Pedersen, E., et al. (2009). "Response to noise from modern wind farms in The Netherlands." *Journal of the Acoustical Society of America*, 126(2): 634–643.

[41] Janssen, S.A., et al. (2011). "A comparison between exposure-response relationships for wind turbine annoyance and annoyance due to other noise sources." *Journal of the Acoustical Society of America*, 130(6): 3746–3753.

[42] Nissenbaum, M.A., Aramini, J.J., and Hanning, C.D. (2012). "Effects of industrial wind turbine noise on sleep and health." *Noise & Health, Sep-Oct*; 14(60): 237–243.

[43] Pierpont, N. *Wind Turbine Syndrome: A Report on a Natural Experiment*. Santa Fe, New Mexico: K-Selected Books, 2009.

[44] Rubin, G.J., Burns, M., and Wessely, S. (2014). "Possible psychological mechanisms for 'wind turbine syndrome.' On the windmills of your mind." *Noise & Health*, 16(69): 116–122.

[45] Tachibana, H., et al. (2014). "Nationwide field measurements of wind turbine noise in Japan." *Noise Control Engineering Journal*, 62(2): 90–101.

[46] Persson, K. and Björkman, M. (1988). "Annoyance due to low frequency noise and the use of the dB(A) scale." *Journal of Sound and Vibration*, 127(3): 491–497.

[47] National Electrical Manufacturers Association (NEMA). *NEMA Standards Publication No. TR 1-2013. Transformers, Step Voltage Regulators and Reactors*. Rosslyn, VA: National Electrical Manufacturers Association, 2014.

[48] Lings, R. *EPRI AC Transmission Line Reference Book – 200 kV and Above*, 3rd edn. Palo Alto, CA: Electric Power Research Institute, 2005.

[49] Clark, W.W. (1991). "Noise exposure from leisure activities: A review." *Journal of the Acoustical Society of America*, 90(1): 175–181.

[50] Kryter, K.D. and Garinther, G.R. (1965). "Auditory effects of acoustic impulses from firearms." *Acta Otolaryngologica Supplement* 211: 1–22.

[51] Johnson, D.L. and Riffle, C. (1982). "Effects of gunfire on hearing level for selected individuals of the Inter-Industry Noise Study." *Journal of the Acoustical Society of America*, 72(4): 1311–1314.

[52] Clark, W.W. (1992). "Hearing: the effects of noise." *Otolaryngology – Head and Neck Surgery*, 106(6): 669–676.

[53] Stewart, M., et al. (2002). "Hearing loss and hearing handicap in users of recreational firearms." *Journal of the American Academy of Audiology*, 13(3): 160–168.

[54] Cowan, J.P. *Handbook of Environmental Acoustics*. New York: Wiley, 1994.

[55] Kardous, C.A. and Morata, T.C. (2010). "Occupational and recreational noise exposures at stock car racing circuits: An exploratory survey of three professional race tracks." *Noise Control Engineering Journal*, 58(1): 54–61.

[56] Rose, A., et al. (2008). "Noise exposure levels in stock car auto racing." *Ear Nose and Throat Journal*, 87(12): 689–692.

[57] Gupta, D. and Vishwakarma, S.K. (1989). "Toy weapons and firecrackers: a source of hearing loss." *Laryngoscope*, 99(3): 330–334.

[58] Altkorn, R, Milkovich, S., and Rider, G. (2005). "Measurement of noise from toys." *Proceedings of Noise-Con 2005*, Minneapolis, MN.

[59] European Committee for Standardization (CEN), Technical Body CEN/TC 52 – Safety of Toys. *EN 71-1:2011+A2:2013 Safety of Toys – Part 1: Mechanical and Physical Properties*, Brussels, 2013.

[60] ASTM International. *ASTM F963-11, Standard Consumer Safety Specification for Toy Safety*. West Conshohocken, PA: ASTM International, 2011.

[61] International Organization for Standardization. *ISO 8124-1:2014- Safety of Toys – Part 1: Safety Aspects Related to Mechanical and Physical Properties*; International Organization for Standardization: Geneva, Switzerland, 2012.

[62] McLaren, S.J., et al. (2014). "Noise producing toys and the efficacy of product standard criteria to protect health and education outcomes." *International Journal of Environmental Research and Public Health*, 11: 47–66.

[63] J. H. Dempster and K. Mackenzie, "The resonance frequency of the external auditory canal in children." *Ear and Hearing*. 11(4) 296–298, (1990).

[64] Kruger, B. (1987). "An update on the external ear resonance in infants and young children." *Ear and Hearing*, 8(6): 333–336.

[65] Katz, A., et al. (1982). "Stereo earphones and hearing loss." *New England Journal of Medicine*, 307(23): 1460–1461.

[66] Keith, S.E., Michaud, D.S. and Chiu, V. (2008). "Evaluation the maximum playback sound levels from portable digital audio players." *Journal of the Acoustical Society of America*, 123(6): 4227–4237.

[67] Fligor. B.J. (2009). "Risk for noise-induced hearing loss from use of portable media players: A summary of evidence through 2008." *Perspectives on Audiology*, 5(1): 10–20.

[68] Williams, W. (2005). "Noise exposure levels from personal stereo use." *International Journal of Audiology*, 44(4): 231–236.

[69] Torre, P. (2008). "Young adults' use and output level settings of personal music systems." *Ear and Hearing*, 29(5): 791–799.

[70] Fligor, B.J. and Cox, L.C. (2004). "Output levels of commercially available portable compact disk players and the potential risk to hearing." *Ear and Hearing*, 25(6): 513–527.

[71] Welch, D., Law, A., and Dirks, K.N. (2014). "Hearing loss with exercise and noise exposure." *Proceedings of ICBEN 2014*, Nara, Japan.

[72] Airo, E., Pekkarinen, J., and Olkinuora, P. (1996). "Listening to music with headphones: An assessment of noise exposure." *Acustica.* 82(6): 885–894.

[73] Dalton, D.S, et al. (2001). "Association of leisure-time noise exposure and hearing loss." *Audiology*, 40(1): 1–9.

[74] Feldmann, J. and Pitten, F.A. (2004). "Effects of low frequency noise on man – a case study." *Noise & Health*, 7(25): 23–28.

[75] McCosker, J.E. (1986). "The Sausalito Hum." *Journal of the Acoustical Society of America*, 80(6): 1853–1854.

[76] French, C.C, et al. (2009). "The "Haunt" project: An attempt to build a "haunted" room by manipulating complex electromagnetic fields and infrasound." *Cortex*, 45: 619–629.

[77] Tandy, V. (2000). "Something in the cellar." *Journal of the Society for Psychical Research*, 64(860): 129–140.

[78] Walford, R.E. (1983). "A classification of environmental "hums" and low frequency tinnitus." *Journal of Low Frequency Noise and Vibration*, 2(2): 60–84.

[79] van den Berg, F. (2001). "Tinnitus as a cause of low frequency noise complaints." *Proceedings of Internoise 2001*, the Hague, Netherlands.

[80] Smith, M., et al. (2013). "Characterization of the 'Ranchlands' Hum"; a Calgary community noise nuisance." *Proceedings of Noise-Con2013*, Denver, CO.

[81] Silber, E.A. and Brown, P.G. *Scientific Research to Characterize and Localize the Windsor Hum: Final Report*. London, ON: University of Western Ontario, 2013.

[82] Novak, C., Charbonneau, J., and D'Angela, P. *Investigation of the Windsor Hum*. Windsor, ON: University of Windsor, 2014.

[83] Mullins, J.H. and Kelly, J.P. (1995). "The mystery of the Taos Hum." *Echoes*, 5(3).

[84] Cowan, J.P. *The Kokomo Hum Investigation*. Acentech Project No. 615411. Cambridge, MA: Acentech Incorporated, 2003.

[85] Deming, D. (2004). "The Hum: an Anomalous Sound Heard Around the World." *Journal of Scientific Exploration*, 18(4): 571–595.

[86] Goodman, S. *Sonic Warfare: Sound, Affect, and the Ecology of Fear*. Cambridge, Massachusetts: MIT Press, 2012.

[87] Pompei, F.J. (1999). "The use of airborne ultrasonics for generating audible sound beams." *Journal of the Audio Engineering Society*, 47(9): 726–730.

[88] Beaman, C. P. and Williams, T.I. (2010). "Earworms (stuck song syndrome): towards a natural history of intrusive thoughts." *British Journal of Psychology*, 101(4): 637–653.

7

Positive Effects of Sound

7.1 Introduction

While most of the effects mentioned up to this point have been negative, there are many positive effects associated with sound. Music has been known throughout history to promote positive effects on people, and entire industries have been established around music psychology and sound therapies meant to assist with the treatments of human ailments. Much research has been performed in this field, but there is also a significant amount of anecdotal information. This is explored in this chapter, along with the importance of natural sounds, leading to a discussion of the concept of soundscapes. The chapter concludes with a discussion of how sound is used to influence people and, although this aspect of sound is included in the discussion of positive effects, it is debatable whether this use of sound is positive or negative.

7.2 Music Psychology

While the word "noise" can be traced to the definitively negative Latin word *nausea*, the word "music" stems from the pleasurable effects related to the Greek Muses. Although the psychological effects related to music have been documented throughout history, the science associated with those effects has been formally documented since the 19th century, most notably beginning with Hermann Helmholtz's *On the Sensations of Tone as a Physiological Basis for the Theory of Music* (first published in 1862, translated from the original German into English and republished in 1954 [1]). Since that time and continuing to the present day, research has been attempting to answer the questions about how and why music affects people so profoundly, while there is no doubt that it does. For thorough discussions on the current state of research in this area, the reader is referred to such texts

The Effects of Sound on People, First Edition. James P. Cowan.
© 2016 John Wiley & Sons, Ltd. Published 2016 by John Wiley & Sons, Ltd.

as the *Handbook of Music Psychology* [2] and the *Handbook of Music and Emotion* [3]. The discussion in this book is limited to the actual effects.

Music's main benefit (or detriment) is the emotional connection created with the individual listener. Until recent times, the emotional connection between music and listeners has been (for the most part) related to the tonal and rhythmic qualities of the music. The exception to this is religious or cultural music that has been associated with emotional experiences due to its association with the spiritual realm through repetition and the inherent beliefs associated with it. Lyrics in contemporary music have added a new dimension to that emotional connection by adding specific meaning to songs and enhancing that connection with people dealing with their own related struggles. This results in subjective responses to music that are a function of personal experiences, making it difficult, if not impossible, to determine general rules or trends for these responses.

The loudness of music, or any sound, can elicit an emotional response. As discussed in earlier chapters, elevated sound levels are most often associated with the negative effects of sound on people. However, certain acoustical environments generate arousal when sound levels are elevated, such as at concerts or rallies [4]. This arousal can be enhanced by generating vibratory sensations and even activating the vestibular system at high, low-frequency sound levels at night clubs and concerts, where typical sound levels have been measured to exceed the threshold of body vibrations in the 100–300 Hz range [5].

Whether or not lyrics are involved, cultural roots affect the types of music to which people become accustomed. Independent of these cultural influences, there are three basic emotions elicited by music – happiness, sadness, and fear. These emotions have been shown to be universal among humans through recent studies with isolated African Mafa tribe members, in which these people were exposed to Western music never heard before to elicit the same emotions as those elicited by the same music on Westerners who were familiar with the music [6]. Another study in the US involved exposing university students to different types of music that elicited the emotions of happiness, sadness, fear, and tension, while physiological measures of cardiac, vascular, dermal, and respiratory functions were monitored. Music associated with sad emotions produced the largest changes in heart rate, blood pressure, skin conductance, and temperature, while music associated with happiness produced the largest changes in respiration [7], clearly demonstrating the physiological changes induced by different types of music.

In a demonstration that the type of music affects emotional states, the same safety film showing industrial accidents was shown to three groups of people – one group being exposed to no music during the film, one group being exposed to non-descript music, and the third group being exposed to dissonant music typically used in horror films. Stress reactions were monitored in each of the subjects and those exposed to the nondescript music exhibited lower stress than those being exposed to no music, while those exposed to the horror film music during the same film exhibited higher stress reactions than those being exposed to no music [8], The amygdala in the brain appears to have the most influence on startle and fear reactions associated with sound [3]. It is one of the emotional centers of the brain, used, in particular, for recognition of danger. The amygdala also regulates memory, which makes it instrumental in associating memories with fears. Since the amygdala is always functioning, it is instrumental in waking us when a sound is associated with danger.

Music in general can have the effect of either distraction or enhancement in task performance, depending on the specific task and personal preference. For example, reaction times during driving can be affected by music. A study in Australia showed that reaction times were enhanced by listening to moderate levels of music but were degraded when listening to higher levels [9]. In this case, 1990s heavy metal dance music was being played in the mid-50s on the dBA scale at the listener for the moderate condition and in the mid-80s on the dBA scale for the higher-level condition. This may not be a universal conclusion, however, as the subjects were university students and subjects of different ages and different music types were not included. Another confounding factor for these studies is personality type. A UK study involving university students evaluated task performance of introverts and extraverts while being exposed to silence, music, and typical office noise [10]. Although task performance for both personality types was negatively affected by music and noise, introverts were more affected than extraverts in each category of exposure. This study also used high-tempo music so, although these results are significant, they can't be generalized to all types of music.

Music types can affect behaviors under different circumstances. This can be used to influence consumer behaviors in different environments, as will be discussed later in this chapter, but general music environments tend to be more conducive to specific types of behaviors. For example, it has been shown in a French study that loud music (88 vs. 72 dB at the center of the room) in bars leads to increased alcohol consumption and faster drinking rates [11].

Other environments heavily influenced by music are exercise facilities. The arousal potential of a generic piece of music (to minimize subjective aspects of the music perception) is based on its tempo and level. A UK study using university students supports the widely held notion that music listened to at higher sound levels and higher tempo (producing higher arousal rates) is preferred for exercising activities, while lower tempo and sound levels are preferred for relaxation [12]. The conditions used in this study included the same generic music samples modified to 80 beats/min and 60 dB at the listener for low arousal and 140 beats/min and 80 dB at the listener for the high arousal. There is general agreement that music accompanying exercise results in increased exertion, a reduction in the rate of perceived exertion, and enhancement of affective states at high levels of work intensity [13].

Although music has many confirmed and widely accepted beneficial effects on people, some effects have been identified in isolated research studies that have not been supported by further research. An extreme example of this is what became to be known as the Mozart effect, stemming from a small California study published in 1993 which morphed into a mass-media educational movement. The original study involved 36 college students who took standardized intelligence tests under three different conditions – after listening to 10 minutes of Mozart's sonata for two pianos in D major, after 10 minutes of listening to a relaxation tape, and after 10 minutes of silence. The results showed an equivalent of eight- to nine-point intelligence quotient (IQ) score increases after listening to the Mozart piece compared with scores under the other conditions; however, the effect only lasted for 10–15 minutes following the exposure [14]. The premise behind this study was extrapolated by the communications media to the general idea that listening to classical music enhances intelligence, a premise that has since then been debated by scientists worldwide. Music therapy work by French physician Alfred Tomatis, reported in his autobiography [15] and publicized through a book trademarking the name of the effect [16], affected curricula worldwide and established an industry of books and audio programs touted to increase intelligence by listening to music. The premise was diffused by 1999 with the publication of studies that clarified these results [17], including comments by the authors of the original 1993 study stating that they never claimed that listening to classical music enhances intelligence and that the effects were limited to spatial-temporal tasks involving mental imagery and temporal ordering. A summary of how this phenomenon began, thrived, and disappeared was published shortly thereafter [18]. This by no means diminishes the potential benefits of music and sound therapies, but merely shows an example of how these benefits can be misinterpreted. Sound therapies, when performed by trained, experienced professionals, can provide highly beneficial results.

7.3 Sound Therapies

Over the past century, music has been associated with variations in the following physiological responses through numerous research studies [2]:

- Heart and pulse rate
- Skin response and temperature
- Respiration rate
- Blood pressure
- Muscular tension/posture
- Blood volume
- Stomach contraction
- Pupillary reflex
- Blood oxygen level
- Hormone secretion.

In this regard, music has been proven through numerous studies to relieve stress [19]. The principle of entrainment can also have a significant effect on one's reaction to music. Entrainment is the process by which energy in one object is projected onto another to the point where the second object is synchronized with the energy of the first. This phenomenon occurs in nature when schools of fish or flocks of geese move in organized patterns. It also occurs in acoustics, manifested through external symptoms of moving with the rhythm of music to internal symptoms of variations in bodily functions and thought patterns. Entrainment is not associated with mechanical resonance. Resonance is a mechanical reaction to physical characteristics associated with sound exposure. Entrainment may have some mechanical components, especially in its manifestations, but it is not frequency-specific as it is not related to the mechanical structure of the object. Entrainment, especially with regard to sound, can also have an emotional component which transcends anything related to mechanical resonance.

The ability of sound energy to manipulate organisms and treat illnesses has been widely documented, both anecdotally and through research programs. Cells have been photographed to be manipulated by exposure to music [20] and medical doctors have documented numerous cases of illness recoveries from sound therapies involving music, voice, and tones [21–23]. As many of these treatments are more anecdotal than scientific in their publications, the reader is referred to these sources with that understanding.

The use of music in treating physical and mental illnesses has been documented for thousands of years. Music therapy has developed into a credible profession since the mid-20th century with the establishment of the American Music Therapy Association (AMTA) in the US, offering a board-certification program from which there are currently over 5,000 members, and later in Italy with the establishment of the World Federation of Music Therapy. A thorough history of the profession of music therapists can be found in the AMTA's latest overview [24]. As noise has been associated with stress-related dis-eases, music is associated with the eases of relieving stress by activating the same systems in the body to result in positive effects. This assumes that the music is pleasing to the individual, however, since taste in music is purely subjective and music can be interpreted as noise depending on the person and circumstance.

Music therapists' work has been documented to ease most physical and mental afflictions, but the key word is "ease" rather than cure. A recent study of cardiac rehabilitation patients showed clear physical (with respect to blood pressure) and psychological benefits from the stress reduction associated with music therapy when it accompanied outpatient rehabilitation exercises [25]. A recent review of 26 studies involving 1,369 participants found a general trend that sedative music could have a beneficial effect for people with coronary heart disease, in terms of lowering systolic blood pressure and heart rate, and also for reducing anxiety for heart attack patients in hospitals [26].

With regard to pain, there are mixed results. A recent review of 51 studies involving 3,663 participants found that listening to pleasing music relieves pain intensity and reduced medication requirements, but the magnitude of that effect is small [27]. Another review of 14 studies shows positive results in pain management, in that music pieces selected by the subject may promote healing through repair mechanisms in the brain with fewer side-effects than those associated with medications [28].

Significant research has also been performed relating music therapy to cancer treatment. A recent review of 30 studies involving 1,891 participants reported positive effects mostly for relieving anxiety and improving mood for cancer patients [29]. Another study review found qualitative improvements for cancer patients in general, with live music by a qualified music therapist being more effective than recorded music [30]. One key aspect of music for patients besides the qualitative improvements is distraction from side-effects, reducing symptom severity for many. There is also some evidence of music therapy boosting the immune system and promoting neurological blocking mechanisms for pain [31]. These are clear examples of the positive psychological effects associated with sound.

7.4 Natural Sources/Soundscapes

The natural environment has long been treasured for its restorative effects from the complexities of urban living [32]. Although urban environments have been considered as being inherently noisy and rural environments inherently quiet, the types of sound sources in these different environments do more to dictate the subjective experience than the actual sound pressure level exposures. It is a valid statement that urban areas have a greater potential to be louder than rural areas because of the many dominant sound sources unique to those environments, but the perceived qualities of those environments are highly subjective.

Some claim that silence is required for a peaceful environment while complete silence can be frightening in many conditions. Although it is generally recognized that an environment in which equivalent noise level ($L_{eq(24)}$) values are less than 45 dBA are acceptable, when speaking in terms of a desire for quiet, the implied aural goal is for "calm" conditions in which noise is not dominant, as defined by the European Environment Agency in their *Good Practice Guide on Quiet Areas* [33].

Each environment has its own soundscape, a term coined by Canadian composer R. Murray Schafer in his 1977 book introducing the subject [34]. The term "soundscape" describes the aural environment associated with a region. It does not imply quiet; it is associated with background sounds inherent to an area. Sound sources not typical for an area are unacceptable components of the soundscape and the goal of a soundscape designer is to preserve the desired aural environment of the community.

The quality of natural quiet has been removed from many areas over the past century with the expansion of transportation networks, with the air travel industry making the most difference, as illustrated in Figure 6.1. Natural quiet in Figure 6.1 is defined by areas not influenced by transportation noise, but true natural quiet is associated with areas devoid of any anthropogenic sound sources. Undisturbed natural environments are being preserved in the US by the National Park Service (NPS), but even they are struggling with man-made sound sources threatening those soundscapes. Trains, buses, cars, motorcycles, snowmobiles, jet skis, military training flights, tourist helicopters, and resource extraction (such as oil, gas, mineral, and timber) negatively affect the soundscape of these areas and efforts continue to reduce activities associated with these sources in protected lands [35,36].

In *Voices of the Wild*, Bernie Krause introduces the field of soundscape ecology, which divides soundscape components into three categories – geophony (natural sources), biophony (sounds from living organisms), and anthropophony (sounds generated by humans) [37]. A concern among soundscape ecologists is that humans are causing changes to soundscapes that are minimizing the availability of natural soundscapes for all species, a point that is illustrated in Figure 6.1 in Chapter 6 for the US.

As background sound pressure levels in some of these areas are below 20 dBA [38,39], preserving the natural quiet has its challenges. The NPS includes soundscape management in their management policies to address these issues by taking action to minimize impacts to the natural soundscapes on NPS properties [40]. To illustrate the value of preserving natural quiet in these settings, visitors to a national park in Spain were recently surveyed about the soundscape and they exhibited a willingness to pay an entrance fee to fund a noise reduction program at the park intended to minimize unnatural sounds [41].

Natural sounds have been shown to promote faster recovery from psychological stress than man-made sounds such as traffic and building ventilation system noise. A recent Swedish study showed that this was true even for lower levels of man-made sounds (by as much as 10 dBA on average; specifically, 50 dBA L_{eq} of bird and fountain sounds resulted in faster recoveries than 40 dBA L_{eq} of man-made ventilation system sound exposure) compared with those of natural sounds [42]. This restorative effect of natural sounds has been recognized by urban planners with their introduction of the quiet-side concept into urban residences, for which a side of the building is facing away from major community noise sources (such as traffic) to create an environment in which the background noise levels are 10–20 dBA quieter than on the side of the building facing the dominant sources. Quiet courtyards can provide quality environments for rest and relaxation, but only in areas with moderate background sound levels. It has been found that when outdoor noise levels exceed an $L_{eq(24)}$ of 60 dBA on the exposed side of the building, the perceived beneficial effect of the quiet side diminishes [43]. The maximum benefit of quiet-side designs occurs when quiet-side background levels are less than an $L_{eq(24)}$ of 48 dBA [44].

It must be considered that the pleasant sounds of birds during the day, insects at night, flowing water through creeks, and ocean waves on beaches can generate sounds into the 50s and 60s on the dBA scale. The qualification of "quiet" must therefore be defined in terms of a minimum of non-natural sounds, rather than an absolute decibel limit. Sounds must also be put into context. The continuous rhythmic sounds of crickets, cicadas, and katydids can be soothing, while the sound of a single mosquito or bee at a lesser sound level and similar frequency range can have the opposite reaction. Placing numerical limits on the soundscape can be impractical as the soundscape concept is based on the inclusion of inherent sounds. The goal is not the absence of sound but the absence of unwanted sound sources that mask or alter the desired acoustic environment.

The urban soundscape is much more complicated than the natural soundscape. Sounds associated with the urban experience form the lifeblood of a city, complete with notable soundmarks (analogous to landmarks) in each area. Soundmarks can be passive or active, with passive soundmarks intentionally added to an area (e.g. fountains and church bells) and active soundmarks generated from human and animal activities. These aspects need to be balanced to result in a pleasant environment. A recent Italian survey showed that the aural environment in historic areas banned from traffic did not turn out to be as pleasant as expected, because isolated sound events were more intrusive with the quieter background [45]. A recent soundscape survey of an historic district in China emphasizes yet another important point, that the acceptability of specific sound sources is subjective and the satisfaction of an aural environment also depends on non-acoustic factors, such as the age of the listener, weather conditions, cultural identity, vitality, sanitation, and landscaping [46]. A natural component to the soundscape is always welcome in any environment, especially through the introduction of birdsong, but preferences tend to change with the age of the individual, leaning towards exclusively natural sounds with increasing age [47].

As mentioned, quiet does not necessarily imply satisfaction with an aural environment. The many typical aural characteristics of an urban environment embody an aspect of the attractiveness of these types of areas. There is therefore no set of acoustic guidelines for ideal soundscapes. The ideal soundscape for any area is unique to that area, driven by the desires of the local population at the time. Soundscapes should therefore be fluid and be dictated by the current inhabitants.

7.5   Using Sound to Influence People

All of the information mentioned thus far supports the premise that sound affects people in physiological and psychological ways; however, these effects have, for the most part, been conscious. Accounts of the influence of sound on people have been passed on throughout history, from the haunting chants of the Sirens in Homer's *Odyssey* in the 8th century BC to religious chants still practiced to this day, sound has been used to influence people's behavior at a local level for thousands of years. The advent of electroacoustic systems in the 20th century created the new potential to influence people with sound on a mass level.

This idea was first introduced by Brigadier General George Owen Squier in the US, with the introduction of electronic mood music in the 1930s. Squier was a pioneer in the telecommunications industry and, later in life, after retiring from a distinguished career in the American Army, he applied his experience to develop a mass communication music service. He presented his patents to the North American Company, which created Wired Radio, Inc. in 1922 to distribute news and dance music for a monthly fee to homes and small businesses. Before his death in 1934, Squier changed the name of the company to Muzak (from a combination of music and Kodak) to give it more of a memorable name. At that point, the Muzak Corporation was researching the effects of sound on people with regard to mood, and the mass production of mood music was expanded into every aspect of commercial life. The idea was to provide aural wallpaper to commercial establishments and workplaces, with nondescript music that created a pleasurable environment in which the music blended in with the atmosphere without creating any distractions. It is, as R. Murray Schafer wrote, "music that is not to be listened to [34]." The story behind Muzak has been chronicled by Joseph Lanza in *Elevator Music* [48], for those interested in more related information.

The premise behind the success of the Muzak Corporation has been researched extensively since the 1930s, and it has been extrapolated into the marketing of products and services on a global level. The general musical characteristics of mode, tempo, pitch, rhythm, harmony, and volume can be varied to produce specific emotional responses [49]. These have been used historically in television and cinema, and have more recently been introduced into the general sales communities. It was initially called "elevator music," because elevators were new inventions in tall buildings at the time and the music was introduced to ease the fears of those using these new devices to quickly ascend and descend between floors.

Another powerful aspect of influential sound is the impact of the lack of it. Silence in an advertisement can have a profound effect, especially when the advertisement includes visual information, such as on television or in the theater. Depending on the context, silence can attract attention to evoke emotions such as intrigue, curiosity, and contemplativeness [50]. Silence can also influence discomfort in an environment. An experiment with 100 students showed that waiting times were underestimated significantly more in conditions where there was music as compared with conditions where there was silence, and three to four times more students left the experiment early after waiting in silence compared with those waiting with music [51]. This was independent of the type of music, suggesting that people are willing to wait longer in conditions with background music than in conditions with silence. Although the distracting nature of music clearly plays a significant role in this, the pleasantness and familiarity with the music also plays a role. This was confirmed in a British study showing that listening to music that people liked or that fit the conditions resulted in longer waiting times on telephone calls [52].

A discussion on the subject of sound's influence on people cannot be complete without mentioning subliminal messaging. Since the 1950s, visual and auditory messages have been added to advertisements with the intention of influencing buying decisions. The exposure of these tactics initially resulted in outrage and fear of unwelcomed manipulations and brainwashing that were not limited to advertising. For the most part, the influential power of these messages was found to be limited, especially in the 1980s after subliminal self-help audio recordings that flooded the market, advertised to improve every aspect of the listener's life, were proved to be ineffective [53].

Since that time, research has revealed more subtle but powerful methods to influence people's buying decisions. A host of studies have showed how music can affect consumer behaviors in different environments. Examples of this premise have been published with regard to retail and occupational environments, with effects based more on emotional response than on brand recognition or purchase intent [54]. Music evoking sad emotions has been shown to influence greeting card purchases more than music evoking happy emotions or no music [55].

Several studies on wine purchases noted that the type of background music in the stores influenced the price of wines sold rather than the volume of sales. One US study showed that patrons tended to buy more expensive bottles of wine when classical music was being played in the store than when top-40 music was being played [56]. In a separate UK study, French and German wines were displayed in a store side by side and patrons bought more of the French wine while French music was playing in the background and more of the German wine while German music was playing in the background [57,58]. This supports the premise of having the music fit the products being sold to maximize sales.

Another aspect of music in shopping environments is time perception. A US study in a clothing store noted that the only effect of varying music in the store was the perception of shopping time. In this case, two types of music were played in the store – background instrumental and foreground original music – in addition to no music for periods of time. Shoppers under 25 years of age reported that they spent more time than expected in the store when exposed to background music and those 25 and older reported the same thing when exposed to foreground original music [59]. A more recent US study by the same researchers revealed that people spent more time shopping in an environment in which unfamiliar music was being played in the background than in environments in which familiar music was being played [60]. This was reported to be associated with the differences attributable to emotional responses related to the music. As people tend to buy more when they spend more time at an establishment, this supports the premise that playing unfamiliar, rather than familiar, music may result in more sales.

The food service industry has had several research studies carried out to determine optimal background conditions for restaurants. In terms of general pitch perception associated with taste, a UK study yielded results showing that sweet and sour foods tend to be associated with higher-pitched sounds, implying that the taste of food may be influenced by the frequency content of background sounds [61].

Music tempo has been shown to be influential in terms of arousal and the length of time patrons spend in restaurants. One US study was set in a restaurant crowded on Friday and Saturday nights, in which the background music tempo was randomly varied by night, with 72 beats/min defining the upper limit of slow tempo music and 92 beats/min defining the lower limit for fast tempo music. The findings revealed that patrons exposed to slow tempo music stayed longer, ate the same amount of food, and consumed more alcoholic beverages than those exposed to fast tempo music [62]. An Australian study investigated the atmosphere created in a restaurant in which different types of music were being played. The results of this study showed that patrons were willing to spend more money in a restaurant when popular, jazz, and classical music were being played in the background [63]. This supports the results of an earlier UK study, which found that no music or easy listening music in the background significantly lowered the perceived value of food and services in the venue [64]. The premise of "fit" is also seen in restaurants, in that music viewed as fitting the environment of the restaurant promotes an expectation of spending more. Research has shown that the music being played in a dining facility can be directly correlated with a person's feeling about the establishment, in that if a person likes the music being played, then that person also likes the atmosphere of the establishment and is likely to return [65]. Yet another aspect that affects patron behavior is the loudness of the music. It has been found that softer music in restaurants results in patrons spending more money than they would in louder establishments [66]. It is also helpful to provide music in waiting areas to shorten the perceived waiting time and minimize patron irritation.

The supermarket is yet another shopping venue studied for the effects of background music on customer behaviors. As for restaurants, studies have shown that the tempo of the music tends to influence the time and money spent at the establishment, with slow tempos resulting in increased revenues [67]. A balance must be struck, however, to provide music that matches the tastes of the majority of patrons, with the understanding that some music genres may offend or irritate some patrons. Ethnic establishments playing ethnic music must tailor the tempo and loudness to the prevailing age group and tempo appropriate to the desired environment [68].

Research has shown that music plays a key role in identifying the image of a consumer establishment. Background music can establish an environment that is perceived as being upbeat or dynamic, aggressive or threatening, and sophisticated or intellectual [69]. Music found to induce pleasure and arousal has been shown to increase consumers' desire to participate in buyer–seller interactions under low and high conditions, but not under moderate conditions [70].

Although the effects of music in retail establishments are consistent, documented occupational effects are not. It is universally accepted that music affects time perception and often makes a job more pleasant, but productivity does not appear to be affected by any type of music, even though perception of increased productivity is often present [71]. This is the case regardless of the complexity of the tasks. Noise, on the other hand, has been associated with negative effects on productivity, as mentioned in Chapter 5.

These principles are now becoming adopted by the business community, with the current emphasis on branding. The branding of a sound signature is becoming just as important to a company's image, if not more so, than its physical logo, most notably with sonic brands associated with companies such as Intel, AT&T, and T-Mobile. Just hearing the few notes associated with these companies immediately brings them to mind. The hope is that these sonic signatures will become earworms [23], which are sonic signatures or parts of songs that stay in our memories for extended periods of time. Also called stuck song syndrome and involuntary musical imagery, it is not completely understood why we lose conscious control of earworms repeating in our minds. What is known is that they mainly depend on an individual's state of mind and personal history, making the phenomenon completely subjective and difficult to generalize [72]. Although earworms have had suggested links with obsessive-compulsive disorder, there has been no proof of that as yet [73].

Sonic logos are carefully developed to be memorable and evoke the kinds of emotions associated with the company brand. For example, the roar of the lion in the MGM sonic logo clearly establishes the image of the company. These have become so important that sound trademarks, or sound marks have been registered with commissions for their legal protections. The first soundmark was the National Broadcasting Company's sequence of three chimes in 1950, followed by others as diverse as Tarzan's yell and the Looney Toons cartoon soundtrack. Full descriptions of the processes involved in developing sonic logos can be found in recent books such as *Sound Business* [74] and *Sonic Branding* [75].

References

[1] Helmholtz, H. *On the Sensations of Tone*. New York: Dover Publications, 1954.

[2] Hodges, D.A. *Handbook of Music Psychology*, 2nd edn. San Antonio: IMR Press, 1996.

[3] Juslin, P.N. and Sloboda, J.A. *Handbook of Music and Emotion: Theory, Research, Applications*. Oxford, UK: Oxford University Press, 2010.

[4] Blesser, B. and Salter, L.R. (2008). "The unexamined rewards for excessive loudness." *Proceedings of ICBEN 2008*, Foxwoods, CT.

[5] Todd, N.P.M. and Cody, F.W. (2000). "Vestibular responses to loud music: A physiological basis of the 'the rock and roll threshold'?" *Journal of the Acoustical Society of America*, 107(1): 496–500.

[6] Fritz, T., et al. (2009). "Universal recognition of three basic emotions in music." *Current Biology*, 19(7): 573–576.

[7] Krumhansl, C.L. (1997). "An exploratory study of musical emotions and psychophysiology." *Canadian Journal of Psychology*, 51(4): 336–352.

[8] Thayer, J.F. and Levenson, R. (1983). "Effects of music on psychophysiological responses to a stressful film." *Psychomusicology*, 3(1): 44–52.

[9] Beh, H.C. and Hirst, R. (1999). "Performance on driving-related tasks during music." *Ergonomics*, 42(8): 1087–1098.

[10] Furnham A. and Strbac L. (2002). "Music is as distracting as noise: the differential distraction of background music and noise on the cognitive test performance of introverts and extraverts." *Ergonomics*, 45(3): 203–217.

[11] Guéguen, N., et al. (2008). "Sound level of environmental music and drinking behavior: A field experiment with beer drinkers." *Alcoholism: Clinical and Experimental Research*, 32(10): 1795–1798.

[12] North, A.C. and Hargreaves, D.J. (2000). "Musical preference during and after relaxation and exercise." *American Journal of Psychology*, 113(1): 43–67.

[13] Karageorghis, C.I. and Terry, P.C. (1997). "The psychophysical effects of music in sport and exercise: A review." *Journal of Sport Behavior*, 20(1): 54–68.

[14] Rauscher, F.H., Shaw, G.L. and Ky, K.N. (1993). "Music and spatial task performance." *Nature*, 365: 611.

[15] Tomatis, A.A. *The Conscious Ear: My Life of Transformation through Listening*. Barrytown, NY: Station Hill Press, 1992.

[16] Campbell, D. *The Mozart Effect*. New York: Avon Books, 1997.

[17] Steele, K.M., et al. (1999). "Prelude or requiem for the 'Mozart effect'?" *Nature*, 400: 827–828.

[18] Bangerter, A. and Heath, C. (2004). "The Mozart effect: Tracking the evolution of a scientific legend." *British Journal of Social Psychology*, 43(4): 605–623.

[19] Pelletier, C. L. (2004). "The effect of music on decreasing arousal due to stress: A meta-analysis." *Journal of Music Therapy*, 41(3): 192–214.

[20] Maman, F. *The Role of Music in the Twenty-First Century*. Malibu, California: Tama-Dõ Press, 1997.

[21] Gaynor, M.L. *The Healing Power of Sound: Recovery from Life-Threatening Illness using Sound, Voice, and Music*. Boston: Shambhala Publications, 1999.

[22] Campbell, D. and Doman, A. *Healing at the Speed of Sound*. New York: Plume, 2011.

[23] Levitin, D.J. *This Is Your Brain on Music: The Science of a Human Obsession*. New York: Plume, 2007.

[24] Davis, W.B., Gfeller, K.E. and Thaut, M.H. *An Introduction to Music Therapy. Theory and Practice. Third Edition*. Silver Spring, Maryland: American Music Therapy Association, 2008.

[25] Mandel, S.E., et al. (2007). "Effects of music therapy on health-related outcomes in cardiac rehabilitation: A randomized controlled trial." *Journal of Music Therapy*, 44(3): 176–197.

[26] Bradt, J., Dileo, C. and Potvin, N. *Music for stress and anxiety reduction in coronary heart disease patients (Review)*.New York: John Wiley & Sons, The Cochrane Collaboration, 2013.

[27] Cepeda, M.S., et al. *Music for Pain Relief (Review)*. New York: John Wiley & Sons, The Cochrane Collaboration, 2010.

[28] Bernatzky, G., et al. (2011). "Emotional foundations of music as a non-pharmacological pain management tool in modern medicine." *Neuroscience & Biobehavioral Reviews*, 35(9): 1989–1999.

[29] Bradt, J., et al. *Music Interventions for Improving Psychological and Physical Outcomes in Cancer Patients (Review)*." New York: John Wiley & Sons, The Cochrane Collaboration, 2011.

[30] Hanser, S.B. (2006). "Music therapy research in adult oncology." *Journal of the Society for Integrative Oncology*, 4(2): 62–66.

[31] Abrams, B. (2001). "Music, Cancer, and Immunity." *Clinical Journal of Oncology Nursing*, 5(5): 222–224.

[32] Hartig, T, Mang, M. and Evans, G.W. (1991). "Restorative effects of natural environment experiences." *Environment and Behavior*, 23(1): 3–26.

[33] Bloomfield, A., et al. *Good Practice Guide on Quiet Areas. EEA Technical Report No. 4/2014*. Luxembourg: European Environment Agency, 2014.

[34] Schafer, R.M. *The Soundscape: Our Sonic Environment and the Tuning of the World*. Rochester, Vermont: Destiny Books, 1994.

[35] Sutherland, L.C. (1999). "Natural quiet: An endangered environment: How to measure, evaluate, and preserve it." *Noise Control Engineering Journal*, 47(3): 82–86.

[36] Miller, N.P. (2008). "U.S. national parks and management of park soundscapes: A review." *Applied Acoustics*, 69(2): 77–92.

[37] Krause, B. *Voices of the Wild: Animal Songs, Human Din, and the Call to Save Natural Soundscapes*. New Haven: Yale University Press, 2015.

[38] National Park Service. *Lake Mead National Recreation Area Acoustical Monitoring 2007–2010*. Boulder City, NV: NPS, 2011.

[39] National Park Service. *Zion National Park. Soundscape Management Plan and Environmental Assessment*. Washington, DC: U.S. Department of Interior, 2010.

[40] National Park Service (NPS). *Management Policies 2006*. Washington, DC: United States Department of Interior, 2006.

[41] Merchan, C.I., Diaz-Balteiro, L. and Soliño, M. (2014). "Noise pollution in national parks: Soundscape and economic valuation." *Landscape and Urban Planning*, 123:1–9.

[42] Alvarsson, J.J.; Wiens, S.; and Nilsson, M.E. (2010). "Stress recovery during exposure to nature sound and environmental noise." *International Journal of Environmental Research and Public Health*, 7: 1036–1046.

[43] Gidlöf-Gunnarsson, A. and Öhrström, E. (2010). "Attractive "quiet" courtyards: A potential modifier of urban residents' responses to road traffic noise?" *International Journal of Environmental Research and Public Health*, 7(9): 3359–3375.

[44] Öhrström, E., et al. (2006). "Effects of road traffic noise and the benefit of access to quietness." *Journal of Sound and Vibration*, 295: 40–59.

[45] Brambilla, G. and Maffei, L. (2010). "Perspective of the soundscape approach as a tool for urban space design." *Noise Control Engineering Journal*, 58(5): 532–539.

[46] Zhou, Z., Dang, J and Jin, H. (2014). "Factors that influence soundscapes in historical areas." *Noise Control Engineering Journal*, 62(2): 60–68.

[47] Yang, W. and Kang, J. (2005). Soundscapes and sound preferences in urban squares: A case study in Sheffield. *Journal of Urban Design*, 10(1): 61–80.

[48] Lanza, J. *Elevator Music: A Surreal History of Muzak, Easy-Listening, and Other Moodsong*. University of Michigan Press, 2004.

[49] Bruner, G.C. (1990). "Music, Mood, and Marketing." *Journal of Marketing*, 54(4): 94–104.

[50] Olson. G.D. (1994). "The sounds of silence: Functions and use of silence in television advertising." *Journal of Advertising Research*, 34: 89–95.

[51] North, A.C. and Hargreaves, D.J. (1999). "Can music move people? The effects of musical complexity and silence on waiting time." *Environment and Behavior*, 31(1): 136–149.

[52] Hargreaves, D.J., McKendrick, J. and North, A.C. (1999). "Music and on-hold waiting time." *British Journal of Psychology*, 90(1): 161–164.

[53] Stroebe, W. (2012). "The subtle power of hidden messages." *Scientific American Mind*, 23(2): 46–51.

[54] Morris, J.D., and Boone, M.A. (1998). "The effects of music on emotional response, brand attitude, and purchase intent in an emotional advertising condition." *Advances in Consumer Research*, 25: 518–526.

[55] Alpert, J.I. and Alpert, M.I. (1989). "Background music as an influence in consumer mood and advertising responses." *Advances in Consumer Research*, 16: 485–491.

[56] Areni, C.S. and Kim, D. (1993). "The influence of background music on shopping behavior: Classical versus top-forty music in a wine store." *Advances in Consumer Research*, 20: 336–340.

[57] North, A.C., Hargreaves, D.J. and McKendrick, J. (1999). "The effect of music on in-store wine selections." *Journal of Applied Psychology*, 84: 271–276.

[58] North, A.C., Hargreaves, D.J. and McKendrick, J. (1997). "In-store music affects product choice. *Nature*, 390: 132.

[59] Yalch, R. and Spangenberg, E. (1990). "Effects of store music on shopping behavior." *The Journal of Services Marketing*, 4(1): 31–39.

[60] Yalch, R.F. and Spangenberg, E.R. (2000). "The effects of music in a retail setting on real and perceived shopping times." *Journal of Business Research*, 49: 139–147.

[61] Crisinel, A.S. and Spence, C. (2010). "A sweet sound? Food names reveal implicit associations between taste and pitch." *Perception*, 39(3): 417–425.

[62] Milliman, R.E. (1986). "The influence of background music on the behavior of restaurant patrons." *Journal of Consumer Research*, 13(2): 286–289.

[63] Wilson, S. (2003). "The effect of music on perceived atmosphere and purchase intentions in a restaurant." *Psychology of Music*, 31(1): 93–112.

[64] North, A.C. and Hargreaves, D.J. (1998). "The effect of music on atmosphere and purchase intentions in a cafeteria." *Journal of Applied Social Psychology*, 28(24): 2254–2273.

[65] North, A.C. and Hargreaves, D.J. (1996). "The effects of music on responses to a dining area." *Journal of Environmental Psychology*, 16: 55–64.

[66] Sullivan, M. (2002). "The impact of pitch, volume and tempo on the atmospheric effects of music." *International Journal of Retail & Distribution Management*, 30(6): 323–330.

[67] Milliman, R.E. (1982). "Using background music to affect the behavior of supermarket shoppers." *Journal of Marketing*, 46: 86–91.

[68] Herrington, J.D. and Capella, L.M. (1996). "Effects of music in service environments: a field study." *The Journal of Services Marketing*, 10(2): 26–41.

[69] North. A.C., Hargreaves, D.J. and McKendrick, J. (2000). "The effects of music on atmosphere and purchase intentions in a band and a bar." *Journal of Applied Social Psychology*, 30(7): 1504–1522.

[70] Dube, L., Chebat, J.C. and Morin, S. (1995). "The effects of background music on consumers' desire to affiliate in buyer seller interactions." *Psychology and Marketing*, 12(4): 305–319.

[71] Newman, R.I., Hunt, D.I. and Rhodes, F. (1996). "The effects of music on employee attitude and productivity in a skateboard factory." *Journal of Applied Psychology*, 50(6): 493–496.

[72] Williamson, V. J., et al. (2011). "How do "earworms" start? Classifying the everyday circumstances of Involuntary Musical Imagery." *Psychology of Music*, 40(3): 259–284.

[73] Beaman, C. P. and Williams, T.I. (2010). "Earworms (stuck song syndrome): Towards a natural history of intrusive thoughts." *British Journal of Psychology*, 101(4): 637–653.

[74] Treasure, J. *Sound Business*. Gloucestershire, UK: Management Books 2000 Ltd, 2011.

[75] Jackson, D.M. *Sonic Branding*. New York: Palgrave Macmillan, 2003.

8

Sound Control and Regulation

8.1 Introduction

The effects of sound can be controlled by physical and regulatory methods. Physical sound control methods involve reducing sound exposure levels to alleviate the unwanted effects or to enhance wanted effects. These methods can be implemented at the sound source, in the path between the source and listener, or at the listener. The physical properties related to sound control are absorption, transmission, diffraction, and cancellation. These properties are described in this chapter, followed by a discussion of trends in regulations related to sound. The chapter concludes with a discussion of priorities in current and planned sound effects research and how the results of this new research will affect the aural environments we will experience in coming years.

8.2 Sound Control Fundamentals

The sound control system is summarized schematically in Figure 8.1. Sound is generated at a source, and the associated sound energy is propagated over a path to be received and interpreted by a listener. If the sound is controlled appropriately at the source, nothing needs to be done in the path or at the listener. If the sound cannot be appropriately controlled at the source, it can be controlled in the path, as is most often the case. If the sound cannot be appropriately controlled at the source or in the path, the sound can be controlled at the listener. The dashed arrows in Figure 8.1 indicate that the system can be more complicated than it seems, with the path potentially affecting sound energy at the source and the listener potentially affecting sound energy in the path. An example of the path affecting the source is a fan in a ducted system, in which sound energy amplified in the duct can affect the sound energy generated by the fan. This can happen in turbine engines with sets of

The Effects of Sound on People, First Edition. James P. Cowan.
© 2016 John Wiley & Sons, Ltd. Published 2016 by John Wiley & Sons, Ltd.

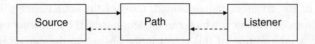

Figure 8.1   The source–path–listener system of sound control

stationary (stator) and rotating blades, where the interaction between sound pressure waves can generate strong tones that can influence the acoustic signature of the system by inducing back-pressure on the rotating blades generating the original sound. An example of the listener affecting the path is especially true for higher-frequency sound, for which wavelengths are smaller than human body components, causing sound to be absorbed, reflected, or diffracted by people. This is one reason why it is best to measure sound pressure levels using meters mounted on tripods at a distance from people or any surfaces that can alter the sound field.

Control measures specific to the source, path, or listener involve general categories of sound properties, so these will be discussed before specific measures. These properties are absorption, transmission control, partial barrier effects, and cancellation.

8.2.1   Absorption

When a sound wave encounters a solid obstacle in its propagation path, the incident sound energy is altered in three ways – some energy is reflected by the partition (as is mentioned in Chapter 1), some energy is absorbed by the partition, and some energy is transmitted through the partition. Dense, smooth materials are highly reflective and porous materials are absorptive. Acoustical absorption is typically described in terms of a unitless absorption coefficient, α, which is the ratio of absorbed to incident sound energy (discussed in Chapter 2). Absorption coefficients range from 0 to 1, with 0 implying no absorption (or total reflection) and 1 implying total absorption (or no reflection). These are ideal limits as some energy is always absorbed and some is always reflected.

Absorption coefficients vary with frequency, most often increasing with increasing frequency for typical porous materials such as fiberglass. The exception to this rule is in the case of discrete frequency absorbers, such as Helmholtz resonators, named for the German physician and physicist Hermann Ludwig Ferdinand von Helmholtz (1821–1894). Helmholtz resonators are characterized by having a relatively small opening with a narrow neck leading to a larger volume chamber, as is the case for typical jugs and beverage bottles. When there is air in the chamber and neck, the volume of air in the chamber acts as a spring attached to the mass of air in the neck, generating a maximum response at the natural frequency associated with this system, as illustrated in Figure 4.1. Assuming the neck is a cylinder (with a circular opening), the resonance frequency (f_h) associated with a Helmholtz resonator is:

$$f_h = (c/\pi)\sqrt{(r/2V)} \qquad (8.1)$$

where c is the speed of sound, r is the radius of the neck opening, and V is the volume of air in the large chamber.

If the walls of the chamber radiate sound, most of the sound energy is concentrated around the resonance frequency, which can be demonstrated by blowing over the top of a bottle. An easy demonstration of this effect is the change in the perceived frequency of radiated sound when a bottle is being filled with liquid. As the liquid fills the chamber, the volume of air in the chamber becomes smaller, thus generating a higher resonance frequency from the acoustic energy generated by the flow of liquid into the chamber.

When the walls of the chamber are dense enough to not radiate sound, the sound energy around the resonance frequency is effectively absorbed. This effect can be enhanced by lining the chamber with porous materials. This principle is being employed in specially designed concrete masonry units with narrow slots opening to larger air cavities, which are used to absorb low frequencies in large indoor spaces requiring durable wall surfaces, such as gymnasiums.

There are two standard single-value parameters that are used most often to describe sound absorption over the human speech frequency range – the sound absorption average (SAA) and the weighted sound absorption coefficient, α_w. SAA is calculated as the arithmetic average of absorption coefficients in the 12 one-third-octave frequency bands between 200 and 2,500 Hz rounded to the closest 0.01, in accordance with ASTM C423 [1]. Versions of ASTM C423 before the year 2000 referenced a noise reduction coefficient (NRC) for this value, which is the arithmetic average of absorption coefficients for the full octave bands between 250 and 2,000 Hz, rounded to the next higher multiple of 0.05. Although SAA has replaced NRC in the standards for some time, NRC values still appear to be used more often in product literature than SAA values. NRC and SAA are based on American standards, while α_w is defined in international standards.

α_w is calculated by matching octave band absorption coefficients between 250 and 4,000 Hz to a standard curve shape included in ISO 11654 [2]. The weighting of this standard curve only affects absorption at 250 and 4,000 Hz to define three shape indicators (L for low, M for medium, and H for high) where absorption exceeds the reference curve by 0.25 or more units. The standard also defines sound absorption classes A–E to categorize α_w values in broader applications. Table 8.1 lists these qualifications for these categories.

Values for all of these absorption ratings < 0.2 are generally considered to be highly reflective and values > 0.7 are considered to be highly absorptive. Although there are many publications that list sound absorption values for different materials, it would be most accurate to use absorption coefficients published by product manufacturers that have been measured using the ASTM or ISO standard methods. In general, hard reflective surfaces

Table 8.1 Sound absorption class ratings according to ISO 11654:1997

Sound absorption class	α_w range
A	0.90–1.00
B	0.80–0.85
C	0.60–0.75
D	0.30–0.55
E	0.15–0.25
Not classified	Less than 0.15

such as painted gypsum and marble have absorption ratings < 0.05 and porous absorptive surfaces such as fiberglass and loose soil have absorption ratings > 0.80.

Absorption is useful for minimizing reflections, most often within an enclosed space, but it can also be useful outdoors when a large (with respect to the wavelength of sound) reflective surface is located near a loud sound source. An acoustically reflective surface near a sound source can reflect up to an additional 3 dB sound pressure level to a listener facing the source and the reflective surface. This is because the listener is exposed to the direct path of sound between the source and listener as well as the extra reflected energy. If the surface is highly reflective, most of the energy will be reflected, resulting in an amount of energy equivalent to two identical sources. As mentioned in Chapter 2, doubling the number of identical sources at the same location results in a 3 dB increase at the listener, compared with the sound pressure level from a single source.

Besides minimizing discrete reflections, acoustical absorption is instrumental in reducing reverberation in a space, as discussed in Chapter 2. Reverberation control is essential for speech intelligibility, most importantly for communications in public spaces and learning environments. As mentioned in Chapter 5, US school design standards call for speech frequency reverberation times to be 0.7 s or less for ideal learning conditions. As reverberation time is directly proportional to the volume of a room and indirectly proportional to the absorption in a room (from equation 2.15), larger rooms require more absorptive treatment than smaller rooms to provide the same reverberation time.

8.2.2 Transmission Control

As absorption controls sound within a room, transmission control addresses sound energy traveling between two enclosed rooms through the perimeter room walls. Transmission control is typically described by the transmission loss (TL) or sound reduction index (R) associated with a sealed partition between two rooms. TL is used in English-speaking countries (mainly in the US) and R is used elsewhere (defined in ISO 10140-2) [3], but they are equivalent measures. TL and R are defined by the following:

$$\text{TL or } R = 10\log\left(W_s/W_r\right) = \text{SPL}_s - \text{SPL}_r + 10\log\left(S/A_r\right) \tag{8.2}$$

where W_s and W_r are the sound powers in the rooms housing the source of sound and the receiver of the sound (after being transmitted through the partition between the two rooms), respectively; SPL_s and SPL_r are the L_{eq} (energy-averaged) values in the source and receiving rooms, respectively; S is the surface area of the common partition between the two rooms; and A_r is the total absorption in the receiving room (defined in equation 2.16, Chapter 2). Note that TL and R are not simply the difference between sound pressure levels in each room; they are also dependent upon the size of the common partition and the absorptive qualities in the room housing the listener. The $10\log(S/A_r)$ term in equation (8.2) generally ranges from –6 to +6 dB, with higher values corresponding to lower absorption in the receiving room. These values also assume the ideal conditions of diffuse sound fields (with sound pressure levels not varying with location) in each room and assume that all of the sound transmitted into the adjacent room is transmitted only through the common partition. These conditions are difficult, if not impossible, to recreate outside of a controlled laboratory environment. In the following discussion, the terms of TL and R are interchangeable.

Unsealed partitions leak sound (similar to unsealed barriers leaking water) and so sealing is critical to sound reduction performance. TL is a measure of the sound reduction associated with a partition, in units of dB, generally increasing with increasing frequency. The exceptions to this are shown in the generic TL spectrum in Figure 8.2.

At low frequencies, the TL of a partition depends on its stiffness and resonance properties. These factors cause significant variation in TL, which is why most homogeneous partitions are not effective at reducing sound energy at low frequencies. Above the resonance frequencies for the partition, typically in the mid-frequency range, TL increases at a rate of 20 × log of the frequency and mass of the partition, known as the mass law region. This translates to a 6 dB increase for each doubling of frequency or mass. The mass law region ends near the bending wave frequency of the partition, at which point there is a sharp drop in TL, as the incident sound energy matches the bending wave frequency of the partition and the energy is more efficiently transmitted in that narrow range, known as coincidence. This can cause problems for homogeneous partitions that have coincidence dips in the speech frequency range, resulting in an unexpected limitation in speech privacy. Half-inch-thick plywood and gypsum board have coincidence dips in the 2,000–3,000 Hz range, which can affect speech privacy. The most effective way to avoid issues associated with coincidence is to use multi-layered non-homogeneous partitions. In that way, each material

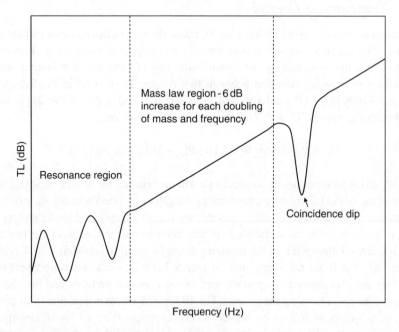

Figure 8.2   Generic sound transmission loss spectrum for a typical homogeneous sealed partition. TL, transmission loss

would have its own coincidence frequency which is different from the others in the full partition, thus having one material compensating for the reduced TL from the coincidence frequencies of the others.

Transmission loss is derived from transmission coefficients, τ, which are unitless ratios of transmitted to incident acoustic energy. Transmission coefficients range from 0 to 1, with 0 implying no energy is transmitted and 1 implying all energy is transmitted. A transmission coefficient of 0 is not possible because some energy will always be transmitted through a partition, but a transmission coefficient of 1 would be the case for any opening (such as an air gap in a partition, an open window or door, or an acoustically transparent material). The transmission coefficient is related to TL in the following way:

$$TL = 10\log(1/\tau) \tag{8.3}$$

As $\log(1) = 0$, a transmission coefficient of 1 implies a TL of 0 dB, which is intuitive since no sound is being reduced by a partition that permits all acoustic energy to be transmitted through it.

As for absorption, there are standard American and international single values representing transmission loss over the human speech frequency range – the sound transmission class (STC) and the weighted sound reduction index (R_w), respectively. STC is specified in ASTM E413 [4] as a single value matched within specified limits to a standardized curve (shown in Figure 8.3) for third-octave band TL data between 125 and 4,000 Hz. R_w is specified in ISO 717 [5] as a single value matched to a similar curve to that for the STC but based on R data ranging from 100 to 3,150 Hz. STC has no units associated with it, although it is based on TL data in dB. R_w data are in units of dB. In each case, the ratings are based on the TL (or R) value at 500 Hz after the curves are appropriately matched, each permitting a summed total deviation below the curve of up to a total of 32 dB, but the STC rating adds the stipulation that the matching curve cannot be more

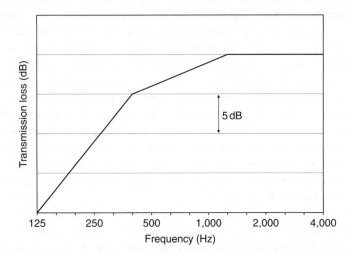

Figure 8.3   Reference curve shape used for sound transmission class determinations

Table 8.2   General speech privacy in adjacent room with single-value partition ratings

STC/R_w range	Speech privacy
0 to 20	No privacy
20 to 40	Some privacy (voices heard in low background levels)
40 to 55	Adequate privacy (raised voices heard in low background levels)
55 to 70	Complete privacy (only high levels of noise heard in low background levels)

than 8 dB above any individual third-octave band TL value (a condition that was also used for older versions of R_w calculations but which has since been dropped). These differences in calculation methods cause STC and R_w values to potentially deviate from each other, although only by slight amounts.

Another point to note about these ratings is that they were developed for human speech privacy determinations and, as such, do not consider low-frequency sounds as much as they consider mid-frequency sounds (as is also shown in Figure 8.3). These ratings are therefore not appropriate for noise reduction ratings associated with unwanted sounds having strong components below 250 Hz.

As for absorption, it would be most accurate to use STC and R_w values published by product manufacturers that have been measured using the relevant ASTM or ISO standard methods. Table 8.2 puts these values into the perspective of speech privacy between fully enclosed rooms, based only on the sound reduction of the common partition between rooms. Other factors must be considered in the full determination of speech privacy and these factors are discussed later in this chapter. The practical upper limit of STC and R_w values is in the 70 range.

A typical wall comprising a single layer of gypsum board on either side of wooden studs spaced 406 mm apart provides an STC/R_w value between 33 and 37. Adding insulation between the studs, increasing the spacing between studs, or replacing the wooden studs with metal studs increases these sound insulation ratings by up to 5 units, and doubling the layers of gypsum boards adds roughly 6 units (from the mass law). Concrete provides more insulation than gypsum, with a typical 152-mm-thick concrete masonry unit block wall providing 43–46 units of reduction. Significantly higher sound insulation values require multi-layered wall designs, with the most effectiveness occurring for double-wall designs (wall sections completely separated by air gaps) or designs including vibration isolation materials to minimize the vibrational energy conducted through solid connections in the walls. In general, the STC or R_w of a partition increases by 5 units with each doubling of air space between double-wall sections, beginning with an air space of 5 cm. Neoprene pads can be effective for isolating wall sections, but resilient clips provide the most effective option for isolating wall sections when air gaps are not desired. Resilient channels are often used for this purpose, but they are also often installed improperly, short-circuiting their intended noise reduction effectiveness.

A key point to bear in mind when a specified sound reduction is required is that STC and R_w values are determined in controlled laboratory environments. Actual installations will invariably be less effective than the laboratory measurements, by up to 5 units if the partitions are completely sealed along their entire perimeters due to unexpected flanking paths

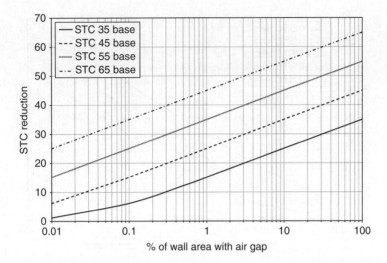

Figure 8.4   Sound transmission class (STC) reduction with air gap percentage for partitions with laboratory STC values ranging from 35 to 65

of sound around the partitions. However, if there are air gaps between the partitions and floors, structural ceilings, or side walls, noise reduction ratings can be severely compromised. Figure 8.4 shows how dramatic this effect can be on sealed walls rated at STC values of 35–65. For example, an air gap covering 0.01% of a wall with a laboratory STC rating of 65 would reduce the effective STC value by 25 units to result in an actual effectiveness of an STC 40 wall. This demonstrates the importance of sealing all air gaps around partitions, including windows and doors, and extending common walls between offices to structural decks whether or not there are dropped ceilings between offices. Typical ceiling tiles in dropped ceilings provide acoustical absorption but little transmission loss, allowing sound to transmit through them to the open spaces above. If these common walls do not extend and seal to structural decks, they will provide the sound reduction more associated with partial barriers than full walls.

Any wall component having a lower STC rating than the main wall will lower its effective STC rating in accordance with the data in Figure 8.5. Windows and doors typically have lower STC ratings than the walls in which they are installed, and Figure 8.5 shows the difference resulting from adding a single component to a wall that is 10, 20, 30, 35, and 40 STC units lower than that of the main wall. For example, if a wall has a window covering 10% of a wall's surface area and the window has an STC rating 20 units less than the wall, the effective STC rating for the composite wall and window combination will be 10 STC units less than the STC rating of the wall alone.

As mentioned earlier, these single-value ratings for both absorption and transmission control are not useful for low-frequency sound, where materials are typically not as effective in sound absorption or reduction. If sound absorption or reduction is desired for sound having dominant components below 250 Hz, the α or TL ratings for partitions in the frequency range of interest would be most appropriate for addressing that.

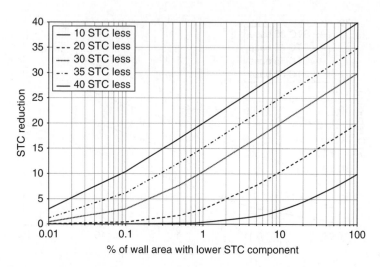

Figure 8.5 Sound transmission class (STC) reduction with lower STC components

There are two variations of these single-value ratings that are also widely used. These are the outdoor–indoor transmission class (OITC) and the impact insulation class (IIC), both of US origin. The OITC is defined in ASTM E1332 [6] as a single-value rating for the sound reduction effectiveness of building exterior walls or wall components. It is based on an assumed outdoor reference A-weighted third-octave band sound spectrum (between 80 and 4,000 Hz) derived from an average of sample transportation noise sources (aircraft, rail, and roadway) given in the standard. The OITC is then calculated by subtracting the TL spectrum of the exterior wall from that reference signal, summing it into a single overall value, and then subtracting that value from a given constant value (100.13). In general, OITC values are lower than STC values by 5–10 units because OITC addresses low-frequency sounds more effectively than does STC.

The IIC is defined in ASTM E989 [7] as a single-value rating for the sound reduction effectiveness of floor/ceiling assemblies with regard to sounds generated by impacts (such as footfalls). For the most part, the determination of IIC is similar to that for the STC but with some reversals. The IIC reference curve extends from 100 to 3,150 Hz and is the reverse of Figure 8.3 in that the drop-off is toward the higher-frequency range (see Figure 8.6). The deviations (for matching sound pressure level data measured from a standard tapping machine on the floor above the receiving room) permitted from this curve are a total of 32 dB with no more than 8 dB in any single frequency band level as for STC, but the deviations of interest are above rather than below the curve. The value of the best matching curve at 500 Hz subtracted from 110 is the IIC.

Minimum acceptable values for multi-dwelling units are typically in the 50 IIC range and IIC values > 65 are considered to be very effective for minimizing footfall noise transmission between floors. Typical hardwood or tile flooring attached directly to a wooden subfloor provides in the IIC range of 30–35 and carpet with padding can provide IIC values > 65.

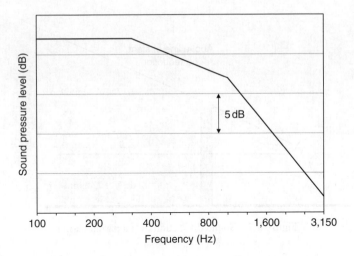

Figure 8.6 Reference curve shape used for impact insulation class (IIC) determinations

8.2.3 Partial Barriers

A partial barrier does not fully enclose a source, so it is open to the air along at least one edge. The sound reduction effectiveness of partial barriers is limited by diffraction, as mentioned in Chapter 1. Figure 8.7 provides a more detailed version of Figure 1.8, showing practical sound pressure level reductions that can be expected from partial barriers. The amount of reduction is dictated by the Fresnel number, N [named for French physicist Augustin-Jean Fresnel (1788–1827)], defined as:

$$N = (2/\lambda)(a+b-d), \tag{8.4}$$

where λ is the wavelength of sound, a is the distance between the source and top of the barrier, b is the distance between the top of the barrier and the listener, and d is the distance between the source and the listener, as shown in Figure 8.7.

Equations have been developed to calculate the relationship between N and the noise reduction provided by a partial barrier and Figure 8.8 is a plot of a simplified version of them. The noise reduction effectiveness of barriers is typically described by the insertion loss (IL), which is the arithmetic difference in sound pressure levels between the conditions with and without the barrier at the same listener location. Barriers provide at least 5 dBA of IL closest to the line-of-sight between the source and the barrier. Their practical upper IL limit is 15–18 dBA, which would occur closest to the barrier in its shadow zone. They also provide minimal IL beyond 100 m from the barrier and are most effective when barriers are either close to the source or listener. When they are within 100 m of the source and listener, the minimum IL occurs when the barrier is equidistant between the source and listener.

The IL values shown in Figures 8.7 and 8.8 assume that the barrier material is solid, flush with the ground, and has an STC or R_w value of at least 20. These conditions ensure that the sound reduction is dominated by the diffraction of sound over and around the

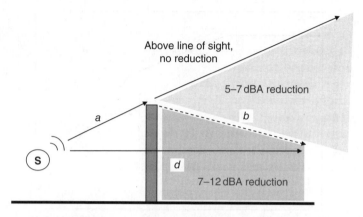

Figure 8.7   Sound diffraction over a partial barrier

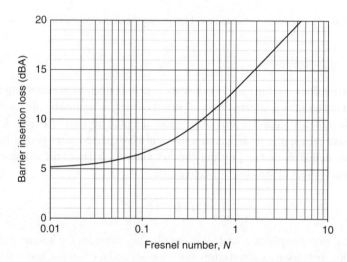

Figure 8.8   Partial barrier insertion loss as a function of Fresnel number (adapted from an equation in Menounou [8]). Note that insertion losses > 15 dBA are often not practical

barrier, so the sound energy being transmitted through the barrier is negligible compared with the amount diffracted around it.

Partial barriers are the typical noise reduction solution chosen by highway and rail agencies to reduce vehicle noise exposures to nearby residences and other noise-sensitive properties. Each agency has its own criteria, mainly in terms of a minimum amount of required noise reduction and cost-effectiveness, for warranting the construction of noise barriers. When one side of a highway or rail line has a noise barrier and the other does not, the highway or rail noise can be amplified on the side without the barrier if the side of the barrier facing the highway or rail line is acoustically reflective. As mentioned earlier in Section 8.2.1, a large reflecting surface close to a source can generate an extra 3 dB of sound pressure level at a listener facing that barrier.

A further complication of highway and rail barriers arises when there are reflective barriers on both sides of a highway or rail line. When the ratio of the distance between the barriers to the height of the barriers is less than 10 to 1, the parallel reflective surfaces can cause an amplification of the highway or rail noise that may be high enough to counter the noise reduction effectiveness expected by diffraction for listeners meant to benefit from the barriers. These issues related to reflective barriers can be solved by using acoustically absorptive materials on the sides of barriers facing the noise sources. Many durable products are available for this kind of use, including porous concrete, perforated metal panels exposing but protecting porous absorptive materials, recycled tires, and walls incorporating vegetation, although vegetation by itself is not effective for use as a sound barrier.

Figure 8.9 shows the common ways in which sound can leak into a room from paths other than through the common partition between rooms (Path A). Path G in Figure 8.9 is the diffracted path when the common partition does not extend and seal to the structural deck of the floor above, as typical dropped ceilings provide minimal sound reduction. This path compromises the sound reduction effectiveness of the common partition more than any other path, giving the illusion of acoustical privacy implied by visual privacy. The other paths are related to air gaps around doors and rigid connections of

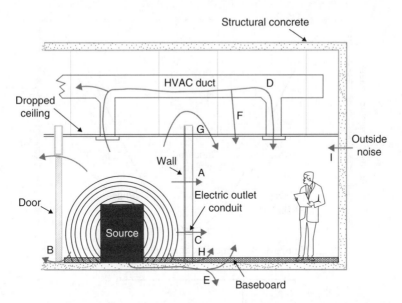

Figure 8.9 Sound leak paths between fully enclosed rooms. Path A – through rigid connections in common partitions; path B – through air gaps around doors; path C – through rigid connections of electrical conduit in walls; path D – through ductwork (HVAC) connected between rooms and radiated through ventilation grilles; path E – through rigid connections between a sound or vibration source and a floor that is shared by the two rooms; path F – through ductwork connected between rooms and radiated through duct walls; path G – sound diffracted over a partition not extended to the structural deck; path H – through rigid connections in baseboard heating systems shared by two rooms; and path I – through exterior partitions

components between rooms. HVAC ductwork common to both rooms can also carry sound from grille to grille or through breakout sound radiating from the ductwork (Paths D and F in the figure).

8.2.4   Cancellation

Sound waves can be cancelled by passive or active methods. Both of these methods are based on the principles of room resonance mentioned in Chapter 1 and illustrated in Figure 8.10, with mirror images (180 degrees out of phase) causing destructive interference at specific frequencies. In the passive case, this destructive interference occurs due to the physical dimensions of a chamber. This is the principle behind the operation of reactive mufflers. A reactive muffler is an expansion chamber, illustrated in its simplest two-dimensional form in Figure 8.11. Sound entering the chamber of length L and cross-sectional area S_1 generates the pressure patterns shown in Figure 8.10 and cancels signals at odd integer multiples of quarter-wavelengths that match L. The larger the ratio between cross-sectional areas of the leading and exiting pipes (S_2) and the main chamber (S_1), the higher the sound reduction at those resonance frequencies, roughly in a $20 \log(S_1/S_2)$, or a 6 dB per doubling of ratio, relationship. Simple expansion chamber mufflers can be very effective at reducing sound levels at specific

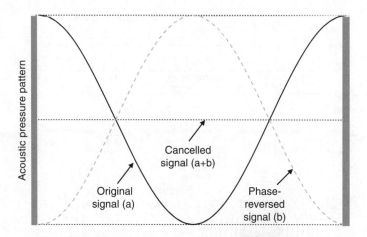

Figure 8.10   Principles of noise cancellation

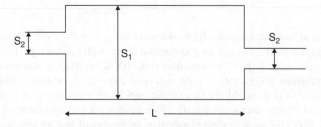

Figure 8.11   Simple expansion chamber muffler design

frequencies, but a more broadband sound reduction can be achieved by using a more complex expansion chamber design, with multiple chambers and holes in the pipes entering and exiting the chambers to generate resonances at a broader range of frequencies.

Another category of mufflers is dissipative, which use acoustically absorptive materials in the interior walls of pipes to reduce sound levels. Dissipative mufflers are generally less effective at noise reduction than reactive mufflers but their advantage in a system is that they provide little, if any, pressure drop through the muffler, which is not the case for reactive mufflers due to the sharp changes in cross-sectional area as flows enter and exit the expansion chamber. The cross-sectional area of flow through pipes does not change with dissipative mufflers. The most effective mufflers offer a combination of reactive and dissipative designs, but installations prohibiting absorptive materials in ducts (such as hospitals and clean rooms) can only use reactive mufflers.

Active noise control creates a destructive interference environment by feeding an electronically generated mirror image of a signal into an enclosed area. This is accomplished by using a system comprising a microphone sensing the original sound fed into a signal processor that reverses the phase of the signal and emits the reversed signal through loudspeakers to cancel the original signal. An additional microphone continually senses how effective that cancellation is and adjusts the signal emitted by the loudspeakers accordingly to maximize the effectiveness of the system. Due to limitations in signal processing speeds, the effectiveness of active noise control systems is limited to low-frequency tonal sound in confined environments [9]. As passive noise control methods are mainly effective for mid and higher frequencies, the ideal noise reduction tool would include active noise control for low frequencies and passive noise control for mid and high frequencies. Ideal uses for active noise control are for reducing low-frequency fan noise in ducted systems. There have been attempts in recent years to replace reactive vehicle mufflers with active noise control mufflers, but the associated costs have hindered mass acceptance of the concept [10].

8.2.5   Control at the Source

The most efficient way to control unwanted sound is to reduce its emissions at the source. This is often a simple matter of equipment maintenance as many noise issues are related to malfunctions or failing or damaged components. The field of predictive maintenance is based on noise and vibration monitoring to detect subtle changes in the noise or vibration signatures that are symptomatic of the early stages of equipment failure. This monitoring can be very effective for equipment with rotating components because they generate tones and associated harmonics directly related to the rotational speed. Any alteration in the standard acoustic signature of this equipment can be indicative of a problem with the equipment that needs to be addressed.

Options for noise control at the source include:

- Equipment maintenance
- Relocating equipment
- Choosing quieter equipment models

- Redesigning equipment
- Removing unnecessary equipment
- Eliminating or minimizing flow turbulence
- Avoiding resonance conditions.

The top five items in this list are self-explanatory. Abrupt changes in pipe or duct direction (through bends or elbows) or cross-sectional area generate turbulence in flows which, in turn, generate noise. Smoothing the flow through curved bends or gradual cross-sectional transitions will minimize turbulence and the noise associated with it. Turbulence is also generated by high flow rates, so minimizing flow rates through pipes and ductwork will minimize flow noise.

Resonance conditions can be mechanical or room-based. Mechanical resonance was introduced in Chapter 4 with regard to body resonance, and is mainly a function of the stiffness and mass of a material. The resonance frequency, at which the vibrational amplitude is maximized (as shown in Figure 4.1), increases with increased stiffness and decreases with increased mass. As Figure 4.1 shows, the goal for operating equipment should be for the resonance frequency of the equipment to be at least one-third of the operating frequency for a diminished response. Therefore, adding mass to a system or reducing the stiffness of the system (using resilient materials) can be used to lower resonance frequencies and thus avoid resonance issues with equipment.

The ideal material for reducing sound transmission and resonance is thus one that has high mass and low stiffness, such as lead or a more environmentally friendly heavy, flexible material such as mass-loaded vinyl. When this kind of material is wrapped around pipes and ducts to control sound radiated from pipe or duct walls (known as breakout noise), it is called lagging material. Effective lagging materials typically provide a breakout noise reduction of 20–30 dBA.

8.2.6   Control in the Path between the Source and Listener

The most common place used to control noise is in the path between the source and listener. This is often the case because noise control at the source or listener is often impractical. Options for noise control in the path include:

- Full source enclosures
- Partial barriers
- Mufflers in ducted systems
- Absorptive treatments
- Vibration isolation
- Active noise control.

Full enclosures can be effective in reducing noise exposures by up to 70 dBA, but operational considerations for most equipment require proper ventilation, thus eliminating the practicality of using a fully sealed enclosure. As mentioned earlier in this chapter, any air gaps can significantly reduce the sound reduction effectiveness of a wall or enclosure. Sound energy will leak from an enclosure through its weakest component, be it an air gap or other component having a much lower transmission loss than the rest of the wall (such as a door or window). Doors, windows, and wall perimeters should be sealed with resilient materials such as silicone caulk and neoprene pads to avoid rigid contact between adjoining surfaces. This not only provides a tight seal to eliminate air gaps but also avoids the transfer of vibrational energy through rigid connections, which is especially important for heavy machinery resting on floors or rooftops. The vibrational energy from the equipment will transmit efficiently through rigid connections to building components and, if those components are rigidly connected to the building's structure, that energy will be transmitted throughout the building. Phantom sources of sound in a building can be generated in this way, with the source potentially being remote from the listener.

Absorptive treatment minimizes reflections between the source and listener, but does not alter the sound level close to the source. This is especially an issue for enclosed sources. Equation (2.16) in Chapter 2 can be used to calculate the total absorption (A) in a room based on the sum of products of absorption coefficients and the surface areas of their associated finishes. When absorptive treatments are added to surfaces in the enclosure, the reverberation time will be affected (in accordance with equation (2.15) in Chapter 2), thus reducing the reverberant sound energy in the space along with the noise level within the space outside of the reverberant field. This reverberant field noise reduction can be described as:

$$NR = 10\log\left(A_2/A_1\right) \tag{8.5}$$

where A_1 is the total absorption in an enclosure before absorptive treatment is added and A_2 is the total absorption after the absorptive treatment is added.

As an example, a room with floor dimensions 20 m × 10 m and a ceiling height of 5 m would have a total surface area of 700 m². If all surfaces in the room have a mid-frequency absorption coefficient rating of 0.05, the total absorption in the room would be 0.05 × 700 = 35 sabins. If absorptive tiles having an absorption coefficient rating of 0.80 are added to cover the entire ceiling, the total absorption of the room with ceiling tiles would be (0.05 × 500) + (0.80 × 200) = 185 sabins. The reverberant field noise reduction resulting from adding the ceiling tiles would then be 10 log(185/35) = 7.2 dB in the mid-frequency range. Practical reverberant field noise reduction is limited to roughly 12 dBA in the reverberant field away from the source but, as absorption varies with frequency, these calculations should be performed consistently for each frequency range involved.

Rotating equipment or components of equipment generate noise based on the shapes and sizes of the rotating components and their rotational speeds. Thinning fan blades and shaping them to minimize turbulence, along with slowing rotational speeds as practical for performance are the key ways to reduce noise generation. Changing rotational speeds can also affect annoyance responses because this will not only lower the sound level but also change the frequency of tones generated by the fan at its blade passage frequency and its harmonics.

8.2.7 Control at the Listener

Noise control at the listener should be the last choice, as this can be the greatest imposition on people. Options for noise control at the listener include:

- Enclosing the listener
- Improving the listener's enclosure insulation
- Relocating the listener
- Using hearing protection devices
- Behavioral therapies
- Adding masking sound.

Hearing protection devices (HPDs) are typically available in two forms – ear plugs and ear muffs. Ear plugs are inserted into the ear canal and ear muffs cover the pinna. Each form of HPD can provide up to 30 dBA of sound reduction to the ear, but they must be used properly to achieve that level of reduction. There are generally three single-value rating methods for HPDs, based on geographical area. The Noise Reduction Rating (NRR) system is used in the US, the Single Number Rating (SNR) system is used in the European Union, and the Sound Level Conversion (SLC_{80}) system is used in the Asia/Pacific region.

Each of these systems is based on relatively complex calculations from regional standards. NRR stems from the US Environmental Protection Agency's regulation in Title 40, Code of Federal Regulations (CFR) Part 211. The NRR value represents the reduction in sound pressure level by the HPD in dBC. For a dBA reduction, 7 units are subtracted from the NRR value. The SNR system stems from ISO 4869-2 [11], and provides three frequency ranges for the ratings – high (H), medium (M), and low (L) – based on correction factors in terms of differences between dBC and dBA values for test signals. The SLC_{80} system stems from the Australia/New Zealand joint standard AS/NZS 1270 [12], which is an estimate of the reduction experienced by 80% of users, with Class designations of 1 to 5. Class 1 is appropriate for reductions up to 13 dB, Class 2 up to 17 dB, Class 3 up to 21 dB, Class 4 up to 25 dB, and Class 5 for reductions greater than 25 dB.

Using plugs and muffs simultaneously will increase the sound reduction to the ear but not by an amount equal to adding the reduction ratings of each HPD. The US Occupational Safety and Health Administration (OSHA) recommends adding 5 dB to the NRR rating of that for the higher of the plug or muff to estimate the total effectiveness of the combination. They also recommend (but don't require) that NRR values are multiplied by 50% (for dBC values or after 7 dB is subtracted for dBA values) as a correction factor for actual workplace conditions [13]. In any case, HPD ratings are based on laboratory measurements and actual performance can vary significantly based on how well the device is sealed in the ear canal (for plugs) or to the side of the head (for muffs).

When noise sources cannot be identified or quieted, as is the case for tinnitus and many hums, cognitive behavior therapy has been used to alleviate the strong negative reactions that may be experienced. This involves relaxation, desensitization, and thought monitoring exercises, and has been reported to provide stress reduction for some trying to cope with noise issues that cannot be solved by traditional methods [14,15].

Acoustic Privacy

Figure 8.9 illustrates the many ways that sound can leak into a room, emphasizing the challenge in achieving acoustic privacy. The concept of architectural design for acoustic privacy was introduced earlier in this chapter in the discussion of transmission loss. The sound reduction attributable to a partition, however, only provides one component toward the attainment of acoustic privacy. As mentioned in Chapter 5, the main acoustic privacy descriptors are related to speech privacy and these are articulation index (AI) and speech intelligibility index (SII). Other related descriptors are privacy index (PI), speech privacy class (SPC), and (loosely) speech transmission index (STI).

The concept of speech privacy was introduced in the 1960s with published results of laboratory and building studies in the US showing that speech privacy is related to speech intelligibility and that increasing background sound levels in a room have a similar effect on speech intelligibility as increasing the transmission loss of a partition between the source and listener. Speech intelligibility was found to be associated with AI, based on a signal-to-noise ratio (a comparison between the speech level and the background sound level) weighted in accordance with human speech frequencies in the 200–6,000 Hz range and peaking in the 800–1,200 Hz range for males [16]. AI varies between 0 and 1, with 0 indicating total speech privacy (or no speech intelligibility) and 1 indicating no speech privacy (or complete speech intelligibility). The SII is a current adaptation of AI, and is defined in the current version of ANSI S3.5 [17]. Older versions of this standard defined AI. SII varies between 0 and 1, as does AI, but due to the calculation methods, SII tends to be slightly higher than AI. The PI is yet another adaptation of a similar parameter, defined in ASTM E1130 [18] as a percentage based on $(1 - AI)$, for use in open-plan environments. The basic concept in the nomenclature is that AI and SII address the level of speech intelligibility (with 0 indicating none and 1 indicating full) while PI addresses the level of privacy (the inverse of intelligibility, with 0% indicating no privacy and 100% indicating full privacy).

The degrees of speech privacy associated with these parameters have been defined in ASTM E1130, with "confidential" speech privacy defined as the condition under which speech cannot be understood and "normal" speech privacy defined as the condition under which concentrated effort would be required to understand speech. Confidential speech privacy is typically only possible in fully enclosed rooms, but normal speech privacy is attainable in an open-plan environment as long as the AI is between 0.05 and 0.20 (or PI is between 80% and 95%). Typical office environments require a sound masking system to elevate background levels, along with partial barriers and reverberation times less than 1 s to achieve normal speech privacy conditions (although room absorption isn't part of these calculations).

The SPC is a new rating system for fully enclosed rooms based on calculations and methods specified in ASTM E2638 [19]. SPC is the sum of the sound insulation associated with the walls surrounding a room and the background sound level inside the room. Rooms having high insulation and low background sound levels would have the same SPC rating as those with low insulation and high background sound levels, as the basis for this rating is the signal-to-noise ratio. The sound insulation is determined by a measurement comparing average sound pressure levels in the source and receiver rooms. Unlike transmission

Table 8.3 Speech privacy rating ranges

AI	SII	PI	SPC	Privacy level
>0.65	>0.75	<35%	<65	Good communications
>0.40	>0.45	<60%	70	No privacy
0.35	0.45	65%	75	Freedom from distraction
0.20	0.27	80%	80	Normal speech privacy
<0.05	<0.10	>95%	85	Confidential speech privacy

AI, articulation index; SII, speech intelligibility index; PI, privacy index; SPC, speech privacy class

loss, this insulation value is simply the difference between average levels in each room; however, there is an option to include TL in the calculation and it is recommended to do so when reverberation times in the receiving room are less than 1.2 s.

There is a distinction made in this standard between the intelligibility and the audibility of speech for rating speech privacy in this way. The threshold of intelligibility is considered to be the condition under which 50% of skilled listeners can just understand some speech, while the threshold of audibility is considered to be the condition under which 50% of skilled listeners can just hear speech. With regard to SPC, the threshold of intelligibility is in the SPC 80 range and the threshold of audibility is in the SPC 90 range. In terms of signal-to-noise ratio, this translates to −16 dB for the threshold of intelligibility and −22 dB for the threshold of audibility for speech sounds between 160 and 5,000 Hz [20]. This means that the threshold of speech intelligibility is 16 dB less than the background sound level and the threshold of audibility is 22 dB lower than the background level.

Table 8.3 provides a summary of all of these privacy ratings along with the meanings of their values with respect to speech privacy.

The STI is mostly applicable to speech intelligibility of electronic sound systems rather than speech privacy. It is based on IEC 60268-16 [21] and was first introduced in the 1980s [22]. Like AI and SII, STI ratings range from 0 to 1 (0 implying no speech intelligibility and 1 implying full speech intelligibility, with privacy ratings similar to those for SII in Table 8.3) and is based on assessing speech intelligibility, but there are several differences between STI and the other speech privacy descriptors mentioned earlier. The main differences are that STI can be directly measured and it is based on monitoring reverberation, distortion, and speech modulation rather than just a signal-to-noise ratio.

8.3 Regulations and Guidelines

Based on the potential health implications discussed in this book, noise regulations have been established in most industrialized countries throughout the world. These regulations have many commonalities, addressing two basic concerns – hearing loss and annoyance or communication. As hearing loss is the only negative physiological effect related to sound that is universally accepted, most industrialized countries have established regulatory limits related to hearing protection and those limits are similar based on the many years of research that have been used to establish those limits. Many sound sources are capable of generating levels that exceed safe hearing thresholds, but only occupational sources and sources related to public exposures can be regulated effectively, leaving responsibilities related to personal equipment at the discretion of each individual choosing to use the equipment.

There has been some effort to educate the public about the potential hazards of high-amplitude sound sources, including labeling of some appliances, but, for the most part, the most potentially damaging recreational sound sources (such as firearms, toys, power tools, personal listening devices, and public performances) are not regulated and have no published warnings associated with them. As noise-induced hearing loss is caused by irreparable (at this juncture in medicine) nerve damage and hearing loss can have debilitating effects on communication, it behooves each person to be knowledgeable of the warning signs related to excessive sound exposures. In general, if verbal communication is impossible in an acoustic environment, the sound levels are most probably exceeding safe limits for at least extended time periods.

8.3.1 Occupational

Occupational noise regulations typically address only potential hazards related to hearing loss. It is universally accepted that a daily 8-hour average exposure level of 85–90 dBA will cause sensorineural hearing loss in the majority of people after 5 years of exposure, with a profound loss occurring after 20 years of exposure. As was discussed in Chapter 4, noise-induced hearing loss is initially manifested as a drop in sensitivity in the 3,000–4,000 Hz range, which is of prime importance to speech recognition. This limit is used as the basis for most occupational noise regulations around the world. The main differences between these regulations are associated with the "exchange rate" used to extrapolate exposure limits to compensate for daily exposures less than 8 hours. The exchange rate is the change in the permissible exposure limit when the exposure time is cut in half, based on the total energy of the exposure rather than the instantaneous value. Most countries agree that the exchange rate should be 3 dB, as would be the case for purely mathematical calculations, because a 3 dB change represents a factor of 2 for energy variations. However, some countries use an exchange rate of 5 dB, resulting in the permissible exposure limits shown in Table 8.4. These kinds of calculations are described using the time-weighted average (TWA), typically based on an 8-hour daily exposure (representing a typical daily work shift).

Table 8.4 Permissible noise exposure limits based on a 5 dB exchange rate and base daily time-weighted average (TWA) of 90 dBA

Duration of daily exposure (hours)	TWA exposure limit (dBA)
8	90
4	95
2	100
1	105
½	110
¼ or less	115

Table 8.4 is the basis for the occupational noise limits in the US, promulgated by the US Department of Labor's Occupational Safety and Health Administration (OSHA) and codified in Title 29 Code of Federal Regulations (CFR) Part 1910.95. When exposures of different levels occur during an 8-hour work shift, satisfaction of equation (8.6) is used to determine whether exposures exceed the prescribed limits.

$$C_1/T_1 + C_2/T_2 + \ldots + C_n/T_n \leq 1 \tag{8.6}$$

where C_n (for $n = 1, 2, \ldots$) is the total daily exposure time to a specific sound level and T_n (for $n = 1, 2, \ldots$) is the permissible noise exposure limit, from Table 8.4, associated with the sound exposure level. For example, if an employee spends 4 hours in a 90 dBA environment, 2 hours in a 95 dBA environment, 1 hour in a 100 dBA environment, and 1 hour in a 70 dBA environment, the relationship in equation (8.6) would be:

$$4/8 + 2/4 + 1/2 + 1/\infty = 1.5$$

As the total is greater than 1, this would constitute an exceedance of the limits. Note that the 70 dBA exposure has no time limit associated with it, so it can either be left out of the equation or included as shown here, where ∞ is infinity and any value divided by infinity is 0.

Most countries use an exchange rate of 3 dB and a baseline 8-hour TWA limit of 85 dBA, resulting in the permissible exposure limits listed in Table 8.5.

Besides the US, only Brazil, Israel, Chile, and Columbia use a 5 dB exchange rate; most others use a 3 dB exchange rate. The US, India, Argentina, Uruguay, and Japan use an 8-hour TWA of 90 dBA as the permissible exposure limit; most others use 85 dBA for that limit, although the European Union and Canada use 87 dBA [23]. The US lowers the limit to an action level of 85 dBA when a standard threshold shift (a change of an average of 10 dB in measured hearing thresholds, compared with a baseline audiogram, at 2,000, 3,000, and 4,000 Hz) is measured during an employee's audiogram. The European Union, through European Directive 2003/10/EC of the European Parliament and of the Council, "On the minimum health and safety requirements regarding the exposure of workers to the risks arising from physical agents (noise), (Seventeenth individual Directive within the meaning of Article 16(1) of Directive 89/391/EEC)," specifies upper and lower action values of 85 and 80 dBA, respectively. Exceeding the action level in the US or upper action value in the

Table 8.5 Permissible noise exposure limits based on a 3 dB exchange rate and base time-weighted average (TWA) of 85 dBA

Duration of daily exposure (hours)	TWA exposure limit (dBA)
8	85
4	88
2	91
1	94
½	97
¼ or less	100

EU requires the implementation of a hearing conservation program, including noise reduction, administrative controls (limiting employee exposures by scheduling), hearing surveillance, employee training, hearing protection, and posting warning signs.

In addition to averaged level limits, most countries having these types of regulations also place limits on impulsive peaks of noise. The US and European Union use 140 dBC as that limit, Australia and New Zealand use an unweighted limit of 140 dB, and the World Health Organization (WHO) recommends impulsive limits of 140 dBA for adults and 120 dBA for children (although these are recreational rather than occupational limits) [24].

8.3.2 Environmental

Environmental noise limits typically address annoyance and communication. The types of sources most often regulated at the national level are transportation-related (roadways, rail lines, and aircraft). Other types of sources regulated on a national level are industrial and, most recently, wind farms. Most other sound sources, if regulated at all, are addressed at the local municipal level. In addition to regulations, there are many unenforceable national standards and guidelines with regard to recommended and preferred sound limits for various indoor and outdoor settings. These are all discussed in the following sections.

National

In the US, the Environmental Protection Agency (EPA) established an Office of Noise Abatement and Control (ONAC) in response to the Noise Control Act of 1972. This agency developed regulations and guidelines for a variety of noise sources until it was defunded in 1981 and that funding has not been restored since that time. The key noise reference document developed by the EPA is known as the levels document [25] from 1974, which established the noise limits identified as "requisite to protect public health and welfare with an adequate margin of safety" listed in Table 8.6.

This document also produced a summary of studies of community reactions associated with outdoor L_{dn} values between 50 and 85 dBA. These values, listed in Table 8.7, have

Table 8.6 US Environmental Protection Agency recommended environmental noise limits [25]

Effect	Limit (dBA)	Type of area
Hearing loss	$L_{eq(24)} \leq 70$	All areas
Outdoor activity interference and annoyance	$L_{dn} \leq 55$	Outdoors in residential areas, farms, and other areas where people spend widely varying amounts of time, and other places where quiet is a basis for use
	$L_{eq(24)} \leq 55$	Outdoor areas where people spend limited amounts of time, such as schoolyards, playgrounds, etc.
Indoor activity interference and annoyance	$L_{dn} \leq 45$	Indoor residential areas
	$L_{eq(24)} \leq 45$	Other indoor areas with human activities such as schools

Table 8.7   US Environmental Protection Agency summary of community reactions
to general outdoor noise levels [25]

Community reaction	Outdoor adjusted L_{dn} (dBA)
Vigorous action	75–85
Widespread threats of legal action	70–78
Widespread complaints	60–72
Sporadic complaints	55–61
No reaction, although noise is noticeable	50–61

been adjusted to consider the many conditions that may affect personal opinions with regard to noise, including such factors as duration of intruding sources, time of year or day of exposure, the tonal or impulsive characteristics of the sources, and personal attitude toward the sources. As these aspects can cause significant variation in results, this information should be taken as general for trends rather than for specific guidelines. This information has been used as a reference for community reaction with increases in noise levels as a step function, with each successive category of community reaction listed in Table 8.7 representing a 5 dBA increase, but this interpretation of these data is not appropriate given the many variables, specific noise levels listed, and the overlapping of noise levels into different categories.

The EPA activities were followed in 1979 by the US Department of Housing and Urban Development (HUD) regulation in Title 24 CFR Part 51B, which defined 65 dBA L_{dn} as the upper limit of acceptable noise environments for housing, based on an interior noise goal of 45 dBA L_{dn}. The US Federal Highway Administration (FHWA), in 1982, published Title 23 CFR Part 772, which established a maximum 1-hour L_{eq} (or L_{10}; however all states except for Minnesota currently use L_{eq}) roadway noise limit of 66 dBA for residential properties. The US Federal Aviation Administration followed in 1989 with Federal Aviation Regulation (FAR) Part 150, establishing a standardized aircraft noise land use compatibility limit of 65 dBA L_{dn} in residential areas. With the only federal industrial noise regulation, the US Federal Energy Regulatory Commission (FERC) instituted Title 18 CFR Part 380 in 1987, limiting natural gas compressor stations to 55 dBA L_{dn} at residential properties.

In 2002, The European Parliament and Council of the European Union passed Directive 2002/49/EC, requiring noise maps to be developed for all municipalities with populations greater than 250,000 near major roadways or rail lines. The prescribed descriptors for these noise maps are L_{den} and L_{night} based on calculations of roadway, rail, aircraft, and industrial noise in each area to assess the number of people that are potentially annoyed or sleep-disturbed throughout Europe, to be updated at least every 5 years to show the extent of the problem. L_{night} is usually the equivalent level corresponding to sounds occurring between 11:00 pm and 7:00 am.

National regulations in general use exterior A-weighted levels in the range of 55–65 dB L_{dn} or L_{den} as their limits. In addition to regulations on community noise, there are guidelines such as those published by the World Health Organization (WHO). The WHO's 1999 community noise guidelines suggest outdoor limits of 55 dBA $L_{eq(16)}$ for serious daytime and evening annoyance, 50 dBA $L_{eq(16)}$ for moderate daytime and evening annoyance, and 45 dBA $L_{eq(8)}$ for sleep disturbance [24]. The sleep disturbance limit has been updated to 40 dBA L_{night} in the WHO update to their 1999 guidelines [26]. In both documents, the indoor sleep disturbance limit is listed as 30 dBA $L_{eq(8)}$. These values are goals that most countries do not use at this time for regulatory limits. In the European Union, only Latvia and the Netherlands use the 40 dBA outdoor L_{night} limit for road traffic noise in new residential communities [26].

That being said, the European Union's Seventh Environmental Action Program of 2013 (in Decision No. 1386/2013/EU) includes an objective that noise levels decrease significantly by 2020, with the goal of "moving closer to WHO recommended levels." The EU defines high noise levels as above 55 dBA L_{den} and 50 dBA L_{night}. With this definition in mind, the European Environment Agency estimates that 125 million people in the EU alone are exposed to high noise levels [27].

Public concern over noise generated by wind turbine farms has prompted recent standards along with legislation dealing specifically with noise from wind farms. There is wide variation among national noise criteria for wind farms and updates are continually being published as new data relating wind farm noise and health are released. Perhaps the most restrictive of wind farm noise criteria are those published in New Zealand (NZS 6808 [28]), for which the limit is 40 dBA $L_{90(10\,min)}$ when background levels (without the wind farm noise) are up to 35 dBA, and 5 dBA above the background when background levels are above 35 dBA. This limit is lowered to 35 dBA under extremely quiet conditions in which background levels are < 25 dBA. Although there is some debate about the appropriateness of an A-weighted descriptor for use in rating wind farm noise (due to the concern over the dominance of low-frequency energy in wind turbines' noise signatures), the premise that these limits are far lower than typical environmental noise limits for other sources is said to compensate for that discrepancy.

There are many noise guidelines dealing with specific sources or types of settings, most notably for schools, health care facilities, and ventilation systems. School guidelines were mentioned in Chapter 5, and similar guidelines to those have been published for health care facilities through the Facility Guidelines Institute (FGI) [29].

A host of descriptors and criteria have been developed for interior ventilation system noise, each referenced to a family of curves that are matched to the noise spectrum in a room and addressing a different aspect of the potential annoyance of the noise signature. The first of these, developed in 1957 and still the most popular, is the noise criterion (NC) set of curves (see Figure 8.12) [30], which are matched with the speech interference level (SIL) discussed in Chapter 5. These were followed by the balanced noise criterion (NCB) curves [31], which extended the frequency range of the NC curves down to 16 Hz (from the original 63 Hz limit), and room criterion (RC) curves [32], which, in addition to specifying an RC value, qualify background sound ratings as neutral or having strong low-, medium-, or high-frequency components in addition to the single-value rating. The most recent is the room noise criterion (RNC) specified in ANSI S12.2 (see Figure 8.13) [33], RNC curves are based on combinations of NC and RC curves to deal more effectively with low-frequency sound while using the more reliable part of the NC curves in the mid-frequency range. Figure 8.14 shows NC and RNC curves on the same chart to illustrate the differences between the two rating methods. As this comparison shows, the curves differ for the most part in the upper frequency range, except at ratings above RNC-45 where there is a significant difference between the two sets of curves at all frequencies.

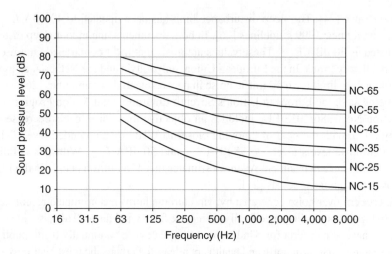

Figure 8.12   Family of noise criterion (NC) curves

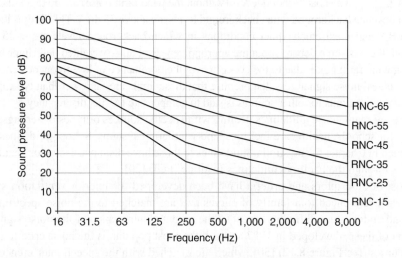

Figure 8.13   Family of room noise criterion (RNC) curves

In each case, the rating is derived by plotting the actual ambient sound pressure level spectrum for a room superimposed with the family of curves of interest. The rating corresponds to the highest curve value that is intersected by the actual room values. Matching single values for each of these designations is used to rate rooms for background sound acceptability. As A-weighted sound pressure levels are roughly 8 units higher than SIL values, NC and RNC values are roughly 8 units less than dBA values. Table 8.8 lists acceptable background noise ratings for different types of common sound-sensitive spaces. The references listed for Table 8.8 provide ranges of acceptable sound levels for more specific rooms in these categories of spaces. Recommendations vary by as much as 10 units between references and the values in this table are the maximum levels listed for each use category.

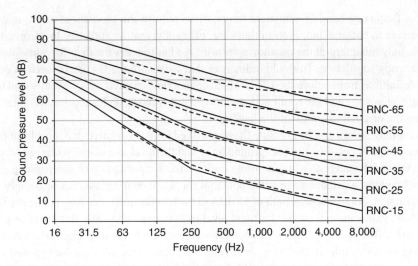

Figure 8.14 Superposition of noise criterion (NC; dashed lines) and room noise criterion (RNC; solid lines) curves (the RNC curves are labeled and the closest dashed curve to each RNC curve is the NC curve of the same value)

Table 8.8 Maximum acceptable indoor background noise limits using common criteria [29,33–35]

Space	dBA SPL	NC or RNC	RC (neutral)
Residences – bedrooms	39	30	30
Residences – living rooms	48	40	35
Hotels/motels	44	35	35
Schools – classrooms	35	30	30
Offices/conference rooms	48	35	35
Offices – open-plan	48	40	40
Health care facilities	44	40	40
Worship spaces	44	35	35
Libraries	48	40	40
Courtrooms	44	35	35

SPL, sound pressure level; NC, noise criterion; RNC, room noise criterion; RC, room criterion

Local

Noise sources not regulated by national codes are in the jurisdiction of local municipalities. Where such regulations exist, they take the form of noise ordinances. Many noise ordinances are qualitative in that they prohibit noise in subjective terms (such as "unnecessary" or "disturbing" noise), making them difficult, if not impossible, to enforce. Some reference "plainly audible" noise, just "audible" noise, or noise that is intrusive for a person with "reasonable sensitivities," qualifications that are entirely dependent on each person's hearing sensitivity. For a noise ordinance to be enforceable, there must be clear quantitative limits that are practical for a soundscape and simple to measure. Complex acoustic criteria, such as frequency band or statistical limits, often require the services of an experienced acoustician to enforce and, even in that case, officers given the authority to enforce these criteria may misinterpret the monitoring results. Add to the mix the fact that some municipalities copy stipulations from old ordinances that have antiquated criteria (such as octave band definitions from before the 1970s which cannot be measured with current sound monitoring equipment) and the result is that enforcement is based on legal prowess rather than serving the original intention of the law.

Quantitative limits can be absolute (having a fixed limit) or relative (having a limit based on a comparison between the background level and the level associated with the source of interest). An advantage to using absolute limits is the simplification of determining a violation; however, a disadvantage is the potential for significant annoyances to comply with the regulation. For example, a limit of 55 dBA in a relatively quiet area may not be sufficient to address annoyances if the background level in that area is 35 dBA and a source generating up to 20 dBA of sound above the background level could comply. This is the advantage of using relative limits in that they are based on a comparison with the background level. Limiting unwanted sources to 5–10 dBA above the background is usually sufficient to protect the public from annoyance; however, it may be difficult to determine the background level, especially if background levels are highly variable.

Background levels can be expressed in terms of L_{90} values as long as those measuring and reporting those values have a clear understanding of their meaning. For sound sources that are not steady (varying by more than ±5 dBA at the listener's location), the most protective ordinance limits would be 5–10 dBA above the L_{90} if the appropriate monitoring equipment and expertise are available. Model noise ordinances are also available that consider these limitations [36].

8.4   Current and Future Research

There is a need for more research in this field to establish more evidence of the links between both the positive and negative effects of sound on people. Except for noise-induced hearing loss, the research to date summarized in this book is not conclusive enough to drive legislation, but only to support guidelines that recommend preventive measures due to evidence suggesting the conclusions mentioned herein. Although there is a clear understanding of noise-induced hearing loss and a clear agreement on the sound levels that cause it, many people continue to be exposed (by both personal choice and otherwise) to sound levels that can damage their sense of hearing. WHO recently published a document in which disability-adjusted life years (DALYs) have been calculated based on noise exposures. DALYs are defined as "the equivalent years of healthy life lost by virtue of being in states of poor health or disability," and WHO estimated there were up to 1.6 million DALYs related to noise exposures in western European countries alone as of 2011 [37]. Research has progressed over the past 20 years and more definitive results on the links between noise exposures and the diseases discussed in this book are forthcoming. Most of this research is being performed in Europe with some studies being performed in the US and other areas of the world.

The greatest effort in reducing noise generation has been and continues to be expended by aircraft agencies around the world, as aircraft noise affects most of the population. Aircraft noise emissions have been reduced by 20–30 dBA between the 1960s and the 1990s, and the goal is to reduce them by another 10 dBA by 2020 through the active efforts of government, university, and aircraft industry research [23]. As roadway noise shares the stage with aircraft noise as the most ubiquitous sources, roadway noise research has been concentrated most recently on developing quieter pavements, but positive results from that research have been limited due to the reduction of noise control effectiveness with pavement wear.

Research is also continuing on the front of the health benefits of sound and music in particular. We have only scratched the surface in this field and more conclusive research is ongoing by such organizations as the American Music Therapy Association and the World Federation of Music Therapy.

Beyond the research, the ongoing challenge is separating the credible from the speculative in this field. It is my hope that this brief discussion will help in that regard.

References

[1] ASTM International. *ASTM C423-09a, Standard Test Method for Sound Absorption and Sound Absorption Coefficients by the Reverberation Room Method*. West Conshohocken, PA: ASTM International, 2009.

[2] International Organization for Standardization. *ISO 11654:1997, Acoustics – Sound Absorbers for Use in Buildings – Rating of Sound Absorption*. Geneva: ISO, 1997.

[3] International Organization for Standardization. *ISO 10140-2:2010, Acoustics – Laboratory Measurement of Sound Insulation of Building Elements – Part 2: Measurement of Airborne Sound Insulation*. Geneva: ISO, 2010.

[4] ASTM International. *ASTM E413-10, Classification for Rating Sound Insulation*. West Conshohocken, PA: ASTM International, 2010.

[5] International Organization for Standardization. *ISO 717-1:2013, Acoustics – Rating of Sound Insulation in Buildings and of Building Elements-Part 1: Airborne Sound Insulation*. Geneva: ISO, 2013.

[6] ASTM International. *ASTM E1332-10a, Standard Classification for Rating Outdoor-Indoor Sound Attenuation*. West Conshohocken, PA: ASTM International, 2010.

[7] ASTM International. *ASTM E989-06 (Reapproved 2012), Standard Classification for Determination of Impact Insulation Class (IIC)*. West Conshohocken, PA: ASTM International, 2012.

[8] Menounou, P. (2001). "A correction to Maekawa's curve for the insertion loss behind barriers." *Journal of the Acoustical Society of America*, 110(4): 1828–1838.

[9] Wise, S. and Leventhall, G. (2010). "Active noise control as a solution to low frequency noise problems." *Journal of Low Frequency Noise, Vibration and Active Control*, 29(2): 129–137.

[10] Hansen, C.H. (2005). "Current and future industrial applications of active noise control." *Noise Control Engineering Journal*, 53(5): 181–196.

[11] International Organization for Standardization. *ISO 4869-2:1994, Acoustics – Hearing Protectors – Part 2: Estimation of Effective A-weighted Sound Pressure Levels when Hearing Protectors are Worn*. Geneva: ISO, 1994.

[12] Standards Australia/Standards New Zealand. *AS/NZS 1270:2002 (R2014), Australian/New Zealand Standard Acoustics—Hearing protectors*, Sydney, Wellington: AS/NZS, 2014.

[13] Occupational Safety and Health Administration. *OSHA eTools, Noise and Hearing Conservation, Appendix IV:C. Methods for Estimating HPD Attenuation*. Washington, DC: US Department of Labor, https://www.osha.gov/dts/osta/otm/noise/hcp/attenuation_estimation.html, 2005.

[14] Leventhall, G., Benton, S. and Robertson, D. (2008). "Coping strategies for low frequency noise." *Journal of Low Frequency Noise, Vibration and Active Control*, 27(1): 35–52.

[15] Leventhall, G., et al. (2012). "Helping sufferers to cope with noise using distance learning cognitive behaviour therapy." *Journal of Low Frequency Noise, Vibration and Active Control*, 31(3): 193–203.

[16] Cavanaugh, W.J., et al. (1962). "Speech privacy in buildings." *Journal of the Acoustical Society of America*, 34(4): 475–492.

[17] Acoustical Society of America. *ANSI S3.5-1997(R2012). American National Standard Methods for Calculation of the Speech Intelligibility Index*. New York: American Institute of Physics, 2012.

[18] ASTM International. *ASTM E1130-08, Standard Test Method for Objective Measurement of Speech Privacy in Open Plan Spaces Using Articulation Index*. West Conshohocken, PA: ASTM International, 2008.

[19] ASTM International. *ASTM E2638-10, Standard Test Method for Objective Measurement of the Speech Privacy Provided by a Closed Room*. West Conshohocken, PA: ASTM International, 2010.

[20] Bradley, J.S. and Gover, B.N. (2010). "A new system of speech privacy criteria in terms of Speech Privacy Class (SPC) values." *Proceedings of 20th International Congress on Acoustics*, Sydney, Australia.

[21] International Electrotechnical Commission. *IEC 60268-16:2011, Sound System Equipment – Part 16: Objective Rating of Speech Intelligibility by Speech Transmission Index*. Geneva: IEC, 2011.

[22] Steeneken, H.J.M. and Houtgast, T. (1980). "A physical method for measuring speech-transmission quality." *Journal of the Acoustical Society of America*, 67(1): 318–326.

[23] National Academy of Engineering, Committee on Technology for a Quieter America. *Technology for a Quieter America*. Washington, DC: The National Academies Press, 2010.

[24] Berglund, B., Lindvall, T., and Schwela, D.H. *Guidelines for Community Noise*. Geneva: World Health Organization, 1999.

[25] USEPA. *Information on Levels of Environmental Noise Requisite to Protect Public Health and Welfare with an Adequate Margin of Safety. Report No. 550/9-74-004*, Arlington, VA: US Environmental Protection Agency, 1974.

[26] World Health Organization. *Night Noise Guidelines for Europe*. Copenhagen: WHO, 2009.

[27] Nugent, C., et al. *Noise in Europe 2014. EEA Technical Report No. 10/2014*. Luxembourg: European Environment Agency, 2014.

[28] Standards New Zealand. *NZS 6808:2010. Acoustics – Wind Farm Noise*. Wellington: Standards New Zealand, 2010.

[29] Sykes, D.M., Tocci, G.C., and Cavanaugh, W.J. *Sound and Vibration Design Guidelines for Health Care Facilities*. Dallas: Facility Guidelines Institute, Acoustics Research Council, 2010.

[30] Beranek, L.L. (1957). "Revised criteria for noise in buildings." *Noise Control*, 3(1):19–27.

[31] Beranek, L.L. (1989). "Balanced noise-criterion (NCB) curves." *Journal of the Acoustical Society of America*, 86(2): 650–664.

[32] Blazier, W.E. (1981). "Revised noise criteria for application in the acoustical design and rating of HVAC systems." *Noise Control Engineering Journal*. 16(2): 64–73 (1981).

[33] Acoustical Society of America. *ANSI S12.2-2008. Criteria for Evaluating Room Noise*. New York: American Institute of Physics, 2008.

[34] Acoustical Society of America. *ANSI S12.60-2010/Part 1. Acoustical Performance Criteria, Design Requirements, and Guidelines for Schools, Part 1: Permanent Schools*. New York: American Institute of Physics, 2010.

[35] American Society of Heating, Refrigerating and Air-Conditioning Engineers. *2015 ASHRAE Handbook – HVAC Applications*, Chapter 48. Atlanta: ASHRAE, 2015.

[36] Cowan, J.P. *Handbook of Environmental Acoustics*. New York: John Wiley & Sons, 1994.

[37] World Health Organization. *Burden of Disease from Environmental Noise: Quantification of Healthy Life Years Lost in Europe*. Copenhagen: WHO, 2011.

Glossary

Absorption – the reduction of sound energy by its conversion into other forms of energy, often heat.

Absorption coefficient (α) – the unitless ratio of absorbed to incident sound energy.

Acoustic impedance – a measure of the resistance imposed on the flow of acoustic energy by a medium, a function of frequency, the density of a medium, and the area through which the flow of energy occurs.

Acoustic reflex – an involuntary contraction of the tensor tympani and stapedius muscles in the middle ear in response to high-level sound exposures.

Acoustics – the science of the properties of sound generation, propagation, and reception.

Acoustic shock injury (ASI) – an effect associated with a sudden, unexpected loud sound exposure, resulting in many symptoms, the most common of which are hyperacusis, tinnitus, headache, and fatigue.

Acoustic trauma – sudden physical injury to the hearing organs of the inner ear by a high-level sound exposure.

Active noise control – the reduction of sound energy by introducing a signal that is 180 degrees out of phase with the original signal.

Amygdala – an area of the brain responsible for emotion and memory, especially related to fear, motivation, and aggression.

Ambient sound level – the sound pressure level associated with all sound sources in an area with no exclusions (contrasted from the background sound level).

Antinode – the maximum pressure associated with a standing wave, resulting from the constructive combination of the original and reflected energy.

Articulation index (AI) – a measure of speech intelligibility, used in the field of acoustics to rate speech privacy and used in the field of audiology to rate speech intelligibility discernment; replaced in updated acoustics standards by the speech intelligibility index (SII).

The Effects of Sound on People, First Edition. James P. Cowan.
© 2016 John Wiley & Sons, Ltd. Published 2016 by John Wiley & Sons, Ltd.

Asymptotic threshold shift (ATS) – a hearing loss limit after continuous exposures to specific levels of sound for at least 8 hours.

Atmospheric absorption – the reduction of sound pressure level by the conversion of sound energy into heat energy at the molecular level in the atmosphere.

Auditory nerve – the eighth cranial nerve (also known as the vestibulocochlear nerve), which carries electrical signals from the cochlea to the brain for interpretation as sound and balance information.

A-weighting – filtering sound levels to match human hearing sensitivity in accordance with the 40 phon equal loudness curve.

Background sound level – the sound pressure level associated with all sound sources in an area in the absence of a specific source of interest.

Balanced noise criterion (NCB) curves – an update for noise criterion (NC) curves, a set of standardized curves that are used to rate the acceptability of background sound pressure levels in rooms extending the curves from a lower frequency limit of 63 to 16 Hz and matched better to the speech interference level (SIL) than NC curves.

Bandwidth – the frequency range over which the sound level response drops off by a factor of 3 dB compared with the level at the peak response frequency.

Beats – pulsing sounds generated by two signals that have dominant tonal acoustic energy close in frequency to each other; the rate of the pulsing sounds is equal to the difference between the dominant frequencies of the two original signals.

Blade passage frequency – a dominant tone associated with rotational equipment, the frequency being a function of the tip speed of the blade multiplied by the number of blades in the equipment.

Bone conduction – sound perception from vibrational energy transmitted through the skull to the inner ear rather than through the ear canal and middle ear.

Brown noise – also known as Brownian noise and red noise, an ideal sound spectrum characterized by sound pressure levels decreasing by a factor of 6 dB per octave with increasing frequency.

Cartilage conduction – sound perception from vibrational energy transmitted from the cartilage of the outer ear, used in a new generation of hearing aids.

Cochlea – the spiral bone-encased section of the inner ear housing the principal organs involved in the transduction of mechanical energy from the vibrations of the middle ear ossicles into electrical signals sent through the auditory nerve to be interpreted by the brain as sound.

Cocktail party effect – the phenomenon of being able to focus attention onto a single sound source in the presence of others of comparable sound level.

Cohort study – a study with groups of people sharing common characteristics or experiences without a medical condition of interest and determining the risk factors involved in contracting that medical condition from exposure to an agent.

Coincidence – the degradation of the transmission loss rating of a homogeneous partition around the bending wave frequency of the partition, also known as the critical frequency.

Community noise equivalent level (CNEL) – the 24-hour energy-averaged sound pressure level, with 5 dB added to sound levels occurring between 7:00 pm and 10:00 pm, and 10 dB added to sound levels occurring between 10:00 pm and 7:00 am; same as L_{den}.

Conductive hearing loss – reduction of hearing sensitivity caused by abnormalities in the outer or middle ear.

Confidence interval (CI) – a statistical parameter of performance designated in terms of a percentage of repeated attempts in which a specific outcome will be within a stated range, typically using 95% in scientific population trend studies.

Critical band – the frequency bandwidth used by the hearing organs in the cochlea for analyzing signals sent to the brain, instrumental in the perception of sound masking.

C-weighting – filtering sound levels to match human hearing sensitivity in accordance with the 90 phon equal loudness curve.

Day–evening–night equivalent level (L_{den}) – the 24-hour energy-averaged sound pressure level, with 5 dB added to sound levels occurring between 7:00 pm and 10:00 pm, and 10 dB added to sound levels occurring between 10:00 pm and 7:00 am; same as CNEL.

Day-night equivalent level (L_{dn}) – the 24-hour energy-averaged sound pressure level, with 10 dB added to sound levels occurring between 10:00 pm and 7:00 am.

Daytime equivalent level (L_{day}) – the 15- or 16-hour energy-averaged sound pressure level, for sound levels occurring between 7:00 am and 10:00 pm or 11:00 pm.

Decibel (dB) – one-tenth of a bel, unit of magnitude (or level), implying a logarithmic (to the base 10) ratio of terms with respect to a reference value; in acoustics, the decibel value is 10 multiplied by the logarithm of a ratio of terms mathematically related to acoustic energy with the reference value related to the threshold of hearing.

Diffraction – the change in propagation characteristics of waves after encountering a partial obstacle that is at least comparable in size to the wavelength of interest; for partial sound barriers, the primary reason for their limited sound reduction effectiveness.

Diffuse field – the region within an enclosure in which sound pressure levels are relatively uniform due to reverberation within the space; same as reverberant field.

Diffusion – the relatively even spreading of energy after waves encounter convex or uneven surfaces.

Direct field – the region in which sound energy from a source is affected by divergence only.

Dissipative muffler – a sound attenuator that uses acoustically absorptive materials to reduce sound pressure levels in a duct or pipe.

Divergence – the even spreading (in a spherical pattern from a point source or cylindrical pattern from a line source) of sound energy over an increasingly larger area with distance from a source.

Earworms – also known as stuck song syndrome and involuntary musical imagery (INMI), musical selections that remain in the conscious mind after they are experienced and are difficult to remove from the conscious mind for an extended period.

Echo – a reflected sound wave from a source arriving at an ear at a delay sufficient to be heard as a distinctly separate sound from the original signal; this delay is typically 50–100 milliseconds.

Electrohypersensitivity (EHS) – a condition for which someone experiences adverse physiological reactions resulting from exposure to typical levels of electromagnetic energy.

Endolymph – a potassium-rich fluid found in the semicircular canals and the scala media (housing the organ of Corti) in the inner ear.

Endolymphatic hydrops – a distension of the membranes separating endolymph and perilymph in the inner ear that can cause the fluids to leak and mix, resulting in hearing loss, tinnitus, feelings of fullness, and vertigo, a condition known as Ménière's disease.

Entrainment – the influence of one energy source on another; with regard to sound and music specifically, the synchronization of an external rhythm with physiological rhythms of a person.

Epidemiology – the science of studying the factors associated with health effects in defined population segments.

Equivalent level (L_{eq}) – an energy-averaged level over a specified time period, or the constant sound pressure level associated with the same amount of energy in the actual signal that is changing in level with time; the reference time period is usually included in parentheses [e.g., $L_{eq(1)}$ refers to a 1-hour L_{eq} value].

Eustachian tube – a channel connecting the air cavity in the middle ear with the rear nasal cavity, instrumental in equalizing pressure between the middle ear and the outside world.

Far field – contrasted to the near field, the region far enough away from a source to be able to record repeatable results, normally at least two wavelengths of interest from a source, where sound pressure levels drop off in accordance with divergence conditions.

Flutter echo – repeated reflections between parallel reflective surfaces.

Frequency – the rate of pressure fluctuations in a sound wave, in units of cycles/second or Hz.

Fundamental frequency – the lowest resonance frequency of a system.

Gray noise – an ideal sound spectrum characterized by the equal-loudness curves included in ISO 226, perceived as equally loud at all frequencies.

Haas effect – also known as the precedence effect, the perception of a single enhanced sound signal in the direction of the first arriving signal when the arrival time difference between two signals is between 1 and 50 milliseconds, and the second signal is less than 15 dB higher in sound pressure level than the first.

Harmonic – a positive integer multiple of the fundamental frequency; the first harmonic corresponds to the fundamental frequency.

Helicotrema – the location at the apex of the cochlea, where the scala vestibuli and scala tympani meet and the lowest frequencies are sensed.

Helmholtz resonator – also known as a volume resonator, an enclosure characterized by a small opening leading into a larger volume chamber, used to absorb or amplify sounds in a limited frequency range associated with the sizes of the chamber and opening.

Hyperacusis – the state or condition of being overly sensitive to normal sounds or having a decreased discomfort threshold for sound.

Hum – a sound heard by a segment of the population with no clearly apparent or measurable (with acoustic instruments) source.

Impact insulation class (IIC) – a single number rating system for the sound reduction effectiveness of floor–ceiling assemblies with regard to impact noises (such as footfalls), for which measured sound pressure levels are matched to a standard curve included in ASTM E989.

Incidence rate ratio (IRR) – a ratio of probabilities of an effect occurring under differing conditions during a specific time period based on counts of those occurrences.

Infrasound – sound energy below 16–20 Hz, only audible at high sound pressure levels.

Inner ear – the segment of the hearing mechanism which converts mechanical energy into electrical energy that is sent to the brain via the auditory nerve for interpretation as sound and balance information; comprising the cochlea and semicircular canals.

Inner hair cells – the ends of nerve fibers in the organ of Corti that are the primary receptors converting their movements into electrical energy sent through the auditory nerve to the brain for interpretation as sound.

Insertion loss (IL) – the difference in sound pressure level, at a specific location, between the conditions with and without a noise reduction device in the path between a sound source and the listener.

Logarithm (log) – in mathematics, an exponent (or power) of a base value.

Lombard effect – the involuntary reaction for which humans (and other species) raise vocal levels and change vocal characteristics when exposed to elevated background sound levels.

Low-frequency sound – generally considered to range from 20 to 200 Hz.

Masking – the properties by which a sound source is inaudible in the presence of another sound source.

Mass law – the portion of the transmission loss spectrum of a partition for which transmission loss increases at a rate of 6 dB per doubling of mass and frequency.

Mechanical resonance – a property of all materials and systems of materials for which there is an increased response when exposed to vibrational energy at the natural or resonance frequency of the system of materials.

Ménière's disease – a distension of the membranes separating endolymph and perilymph in the inner ear that can cause the fluids to leak and mix, resulting in hearing loss, tinnitus, feelings of fullness, and vertigo, also known as endolymphatic hydrops.

Middle ear – the segment of the hearing mechanism between the outer and inner ear segments, including three bones that conduct mechanical energy from the vibration of the tympanic membrane to the entrance of the cochlea at the oval window.

Misophonia – the hatred of sound, a disorder for which a negative psychological response is elicited by exposure to specific sound sources independent of the sound pressure level.

Mozart effect – the premise that exposure to Mozart's music can enhance short-term intelligence, stemming from a small study in 1993 and debunked by 1999.

Natural frequency – the resonance frequency of a system, the frequency at which a maximum response will occur from a vibratory exposure; the frequency at which a system will vibrate if disturbed by a discrete force.

Near field – the region within two wavelengths of interest from a source, where repeatable sound pressure level measurements are not possible due to large pressure fluctuations and no clear drop-off rate.

Night-time equivalent level (L_{night}) – the 8- or 9-hour energy-averaged sound pressure level, for sound levels occurring between 10:00 pm or 11:00 pm and 7:00 am.

Node – the minimum pressure associated with a standing wave, resulting from the destructive (cancellation) combination of the original and reflected energy.

Noise – a subjective, negative interpretation of sound energy.

Noise criterion (NC) curves – the first generation of a set of standardized curves that are used to rate the acceptability of background sound pressure levels in rooms.

Noise-induced hearing loss (NIHL) – sensorineural hearing loss caused by exposure to high noise levels.

Noise reduction coefficient (NRC) – the arithmetic average of sound absorption coefficients at 250, 500, 1,000, and 2,000 Hz; a single-value representation of acoustical absorption over the human speech frequency range used in the US.

Noise reduction rating (NRR) – used in the US, a single-value representing the sound reduction effectiveness of a hearing protection device.

Nosocusis (or nosoacusis) – the loss of hearing sensitivity caused by medical conditions.

Nystagmus – the involuntary movement of the eyes in response to an external stimulus.

Octave band – standardized constant-percentage frequency segments used to evaluate acoustic signatures in more detail than overall levels that offer a single decibel value to cover the entire audible spectrum, designated by geometric mean center frequencies that represent the peak spectral responses for each band; each subsequent octave band center frequency is twice the preceding one, with the eight most common bands being centered at 63, 125, 250, 500, 1,000, 2,000, 4,000, and 8,000 Hz.

Odds ratio (OR) – the probability of an event occurring in one group divided by the probability of the event not occurring or occurring in another group.

Organ of Corti – the main transduction center of the cochlea housing the inner and outer hair cells that convert mechanical energy from vibration-induced fluid motion into electrical energy that is transmitted through the auditory nerve to the brain.

Ossicles – the three small bones (malleus, incus, and stapes) in the middle ear that transmit and amplify mechanical energy related to sound exposure from the tympanic membrane to the oval window.

Otoacoustic emission – sounds generated from within the inner ear that can be measured to determine whether abnormalities exist in cochlear functions.

Otosclerosis – a disease causing the middle ear ossicles to become stiffened (most commonly for the stapes to become rigidly affixed to the oval window) and thereby reducing the transmission efficiency of mechanical energy (associated with sound) to the inner ear.

Outer ear – the segment of the hearing mechanism from the pinna (exposed to the outside world) through the ear canal to the tympanic membrane.

Outer hair cells – the ends of nerve fibers in the organ of Corti that are the secondary receptors converting their movements into electrical energy sent through the auditory nerve to the brain for interpretation as sound; outer hair cells are mostly efferent (receiving feedback information from the brain) and have the ability to expand and contract, unlike inner hair cells.

Outdoor–indoor transmission class (OITC) – a single-value rating for the sound reduction effectiveness of a sealed exterior wall or wall component, based on subtracting the transmission loss spectrum from an A-weighted reference spectrum based on average transportation source noise, specified in ASTM Standard E1332 in the US.

Oval window – the membrane at the connection between the stapes of the middle ear and the cochlea.

Overtone – a positive-integer multiple of the fundamental frequency, the first overtone corresponding to the second harmonic or twice the fundamental frequency.

Percentile level (L_n) – the sound pressure level exceeded n percent of the time period of interest.

Perilymph – a sodium-rich fluid found in the scala vestibuli and scala tympani sections of the cochlea, as well as the region surrounding the semicircular canals.

Period – the time between repeating cycles of pressure variations in a pure tone; the reciprocal of frequency.

Permanent threshold shift (PTS) – irreversible noise-induced sensorineural hearing loss.

Phon – a unit of perceived loudness in accordance with the equal-loudness curves published in ISO 226, based on the frequency sensitivity of human hearing at varying sound pressure levels.

Pink noise – an ideal sound spectrum characterized by sound pressure levels having an equal amount of energy per octave band, named for its light analogy as having higher energy at lower frequencies on a linear frequency scale.

Precedence effect – the perception of a single enhanced sound signal in the direction of the first arriving signal when the arrival time difference between two signals is between 1 and 50 milliseconds, and the second signal is less than 15 dB higher in sound pressure level than the first, also known as the Haas effect.

Presbycusis – the loss of hearing sensitivity caused by aging.

Privacy index (PI) – a measure of speech privacy, defined as a percentage and calculated as the articulation index (AI) subtracted from 1.

Pure tone – a sound pressure wave with dominant energy at a single frequency.

Reactive muffler – a sound attenuator that uses expansion chambers and/or tuned resonators to reduce sound pressure levels in a duct or pipe.

Recruitment – a rapid growth in the ability to hear loud sounds normally while having a hearing threshold shift and thus not having the ability to hear quiet sounds, thought to be a function of which hair cells are damaged in the cochlea.

Red noise – also known as Brownian noise and brown noise, an ideal sound spectrum characterized by sound pressure levels decreasing by a factor of 6 dB per octave with increasing frequency, named for its light analogy as having higher energy at lower frequencies.

Reflection – a change of direction in the path of a sound wave resulting from an interaction with a non-porous material that is larger in size than the wavelength of incident sound energy or a sudden significant change in acoustic impedance.

Refraction – a change of direction in the path of a sound wave resulting from a change in medium conditions that cause the speed of sound to change.

Relative risk (RR) – the ratio of probabilities of a specific occurrence in a group of people exposed to an agent to that occurrence happening in a group of people not exposed to the agent.

Resonance – an increased response to an external stimulus at specific frequencies associated with the density, shape, or stiffness of a system.

Resonance frequency – the natural frequency of a system, the frequency at which a maximum response will occur from a vibratory exposure; the frequency at which a system will vibrate if disturbed by a discrete force.

Reverberant field – the region within an enclosure in which sound pressure levels are relatively uniform due to reverberation within the space; same as diffuse field.

Reverberant field noise reduction – the reduction in sound pressure level in the reverberant field of a space resulting from adding acoustically absorptive materials in that space.

Reverberation – the amplification of sound pressure levels in an enclosed space resulting from multiple reflections off room surfaces.

Reverberation time – the time, in seconds, for the sound pressure level in the reverberant field to decay by a factor of 60 dB.

Room criterion (RC) curves – a set of standardized curves that are used to rate the acceptability of background sound pressure levels in rooms, including qualifications for neutral or strong components in low (rumble) or high (hiss) frequency ranges.

Room noise criterion (RNC) curves – a recent set of standardized curves that are used to rate the acceptability of background sound pressure levels in rooms, based on a combination of the best aspects of noise criterion (NC) and room criterion (RC) curves.

Room resonance – the generation of standing waves in an enclosure in which the distance between parallel reflective surfaces matches integer multiples of half-wavelengths of sounds.

Round window – a membrane on the outer wall of the cochlea (below the oval window) that flexes to equalize the pressure created by movement of the oval window and fluids inside the cochlea.

Semicircular canals – three liquid-filled curved tubes oriented 90 degrees with respect to each other, located directly above the cochlea, and lined with hair cell nerve endings that respond to motions of the head to send electrical signals through the auditory nerve to be interpreted by the brain as balance information.

Sensorineural hearing loss – reduction of hearing sensitivity caused by abnormalities in the inner ear or neural pathway from the inner ear to the brain.

Shadow zone – the region acoustically shielded by a partial barrier, receiving between 5 and 15 dBA of sound pressure level reduction from diffraction.

Single number rating (SNR) – used in Europe, a standard method for rating the noise reduction effectiveness of hearing protection devices, providing ratings in three frequency ranges (high, medium, and low).

Sociocusis (or socioacusis) – the loss of hearing sensitivity caused by non-occupational noise sources, such as environmental or recreational exposures.

Sone – a unit of loudness correlated with phons, for which 1 sone = 40 phons and each doubling of perceived loudness (10 phon increase) above 40 phons corresponds to doubling of the number of sones.

Sound – energy capable of stimulating a hearing mechanism.

Sound absorption average (SAA) – an updated version of NRC, the arithmetic average of absorption coefficients at the third-octave bands from 200 to 2,500 Hz, rounded to the nearest .01.

Sound exposure level (SEL) – also single event level, the sound pressure level representing the total acoustic energy of a discrete event, such as a train pass-by or an aircraft overflight, compressed into a 1-second time interval.

Sound level conversion (SLC$_{80}$) – used in Australia and New Zealand, a standard method for rating the noise reduction effectiveness of hearing protection devices, based on estimated reductions experienced by 80% of users.

Sound power level (L_w or PWL) – the characteristic sound level (in decibels) associated with a source independent of distance, mathematically defined as 10 multiplied by the logarithmic ratio of the power (in Watts) divided by a reference power of 1×10^{-12} Watts.

Sound pressure level (L_p or SPL) – the sound level experienced by a listener and measured by a sound level meter, mathematically defined as 20 multiplied by the logarithmic ratio of the pressure (in microPascals [µPa]) divided by a reference pressure associated with the threshold of hearing (20 µPa).

Sound reduction index (R) – a measure of the sound reduction effectiveness of a partition, in decibels, defined mathematically as 10 multiplied by the logarithm of the ratio of sound powers on either side of a partition between enclosed rooms; same as transmission loss (TL) used mainly in the US.

Soundscape – the aural environment associated with a region, comprised of the background sources that define that environment as unique.

Sound transmission class (STC) – a laboratory-based, unitless, single-value rating for the sound reduction effectiveness of a sealed partition, based on matching the transmission loss spectrum of the partition to a standard curve within tolerances specified in ASTM Standard E413 in the US.

Spectrum – the acoustic signature of a source, a chart showing sound level on the vertical axis against frequency on the horizontal axis.

Speech intelligibility index (SII) – the updated version of articulation index, a rating of speech intelligibility based on a frequency-weighted speech-to-noise ratio, typically used for speech privacy assessments in offices.

Speech interference level (SIL) – the arithmetic average of background sound pressure levels at 500, 1,000, 2,000, and 4,000 Hz, used in rating the acceptability of an environment for comfortable communication.

Speech privacy class (SPC) – a rating of speech intelligibility between enclosed rooms based on the sum of the noise reduction effectiveness of the common partition between rooms and the background sound pressure level in the listener's room.

Speech transmission index (STI) – a rating of speech intelligibility used for lecture and performance spaces, taking into account the reverberation within the space and measurable with instruments, unlike articulation index and speech intelligibility index.

Standing wave – a room resonance phenomenon in which sound waves between parallel reflective surfaces generate pressure patterns that stand still at frequencies associated with integer multiples of half-wavelengths of sound matching the distance between the parallel reflective surfaces.

Stapedius muscle – the smallest muscle in the human body, supporting the stapes in the middle ear; this muscle contracts when a person is exposed to high sound pressure levels to stiffen the stapes and thus reduce the vibrational energy being conducted by the middle ear ossicles to the cochlea as a protective measure.

Temporary threshold shift (TTS) – a loss of hearing sensitivity, usually due to high-level sound exposure, that is followed by a full recovery of hearing sensitivity after a relatively short period of time.

Tensor tympani muscle – a muscle in the middle ear connecting the malleus to the tympanic membrane; this muscle contracts when a person is startled, eating, and speaking to stiffen the eardrum and thus reduce the vibrational energy being conducted by the middle ear ossicles to the cochlea as a protective measure.

Tinnitus – a condition for which sounds are heard that are not generated by external sources.

Tonic tensor tympani syndrome (TTTS) – a condition for which the tensor tympani muscle becomes uncontrollably active, thought to be a result of acoustic shock injury and a possible cause of hyperacusis.

Transmission coefficient (τ) – the unitless ratio of transmitted to incident sound energy, used as the basis for the calculation of transmission loss.

Transmission loss (TL) – a measure of the sound reduction effectiveness of a partition, in decibels, defined mathematically as 10 multiplied by the logarithm of the reciprocal of the transmission coefficient; same as sound reduction index (R) used outside the US.

Tullio phenomenon – a condition for which sounds of varying intensities cause episodes of vertigo, nausea, and nystagmus, thought to be associated with an abnormal third membrane or opening in the inner ear (in addition to the oval and round windows) that causes an imbalance of the fluids in the semicircular canals.

Tympanic membrane – commonly known as the eardrum, the connection between the outer ear and the middle ear.

Ultrasound – sound energy above 20,000 Hz, only audible at high sound pressure levels.

Vibroacoustic disease – a medical condition allegedly caused by long-term exposure to high levels of low-frequency sound, based on a study of aircraft mechanics in Portugal but not supported by other studies.

Wavelength – the physical distance between repeating cycles of pressure variations in a pure tone.

Weighted sound absorption coefficient (α_w) – the international version of the noise reduction coefficient, calculated by matching octave band absorption coefficients between 250 and 4,000 Hz to a standard curve shape included in ISO Standard 11654.

Weighted sound reduction index (R_w) – similar to sound transmission class and used outside the US, a laboratory-based, single-value rating for the sound reduction effectiveness of a sealed partition, based on matching the transmission loss spectrum of the partition to a standard curve within tolerances specified in ISO Standard 717; the R_w covers a slightly different frequency range than STC (100–3,150 Hz rather than 125–4,000 Hz) and is in units of decibels.

Weighting networks – standardized mathematical adjustments to spectra used to account for frequency sensitivities associated with the unit of interest.

White noise – an ideal sound spectrum characterized by sound pressure levels having an equal amount of energy for all audible frequencies on a linear scale, named for its light analogy as having equal energy at all visible frequencies.

Wind turbine syndrome – also known as visceral vibratory vestibular disturbance (VVVD), a multi-symptom medical condition coined by a book of the same name and based on a single limited study in the US, allegedly related to low-frequency noise exposures from wind turbine generators.

Index

The Effects of Sound on People, First Edition. James P. Cowan.
© 2016 John Wiley & Sons, Ltd. Published 2016 by John Wiley & Sons, Ltd.